Plant Science for Sustainable Agriculture and Rural Development

Plant Science for Sustainable Agriculture and Rural Development

Editors
U.N. Bhale
V.S. Sawant
Department of Botany,
Arts, Science and Commerce College,
Naldurg – 413 602
Tal. Tuljapur, Dist. Osmanabad,
Maharashtra, India

2012
DAYA PUBLISHING HOUSE®
New Delhi - 110 002

ISBN 978-81-7035-784-1

Published by : **Daya Publishing House®**
4760-61/23, Ansari Road, Darya Ganj,
New Delhi - 110 002
Phone: 23245578, 23244987
Fax: (011) 23260116
e-mail : dayabooks@vsnl.com
website : www.dayabooks.com

Laser Typesetting : **Classic Computer Services**
Delhi - 110 035

Printed at : **Chawla Offset Printers**
Delhi - 110 052

PRINTED IN INDIA

Preface

This proceeding is an outcome of the conference on "Frontiers of Plant Sciences for Sustainable Agriculture and Rural Development" held on 28th and 29th January 2011 at Department of Botany, Arts, Science and Commerce College, Naldurg, Tq Tuljapur, Dist. Osmanabad affiliated to Dr. Babasaheb Ambedkar Marathwada University Aurangabad (MS). University Grants Commission (UGC), WRO, Pune provided the funds to hold this conference. The conference received an overwhelming response from all over the country with the submission of 157 research abstracts along with invited talks. Out of these abstracts, we have selected 39 full length manuscript for proceeding.

The main objectives of the conference were to create awareness about the emerging trends in agriculture designing, developing new economic process to farmers and provide a platform for forthcoming researchers. We have attempted to compile the invaluable and carefully selected contributions of plant biologists in this consolidation of proceedings of the conference with all ingenuity and alertness.

The editors are grateful to the Principal, Dr. S.D. Peshwe who gave support to host the conference. We are grateful to UGC, WRO, Pune for financial assistance. We are sincerely thanks to all contributors and those directly or indirectly have rendered their help to bring out this proceeding. Our special thanks to Prof. B.F. Rodrigues and Prof H.C. Lakshman for helping us in various ways during the preparation of proceedings. Lastly, our sincere thanks to publisher who provides an opportunity to publish conference proceeding volume.

U.N. Bhale

V.S. Sawant

Foreword

Sustainable agriculture and rural development are significantly related matters. There is need to examine the basic issues in our living and an urgency to call a halt to all crop-production practices which devastates our natural resources and which infringe on the health and happiness of the people. The need of the hour is to test and adopt good tenets of production systems in tune with nature leaving behind gradually the harmful inputs. One of the striking features of sustainable agriculture is management of natural resources before they are completely obliviated so as to take stock of the status of each asset and use them in consonance with the natural endowments in a balanced manner. This calls for attitudinal changes and relearning of traditional practices with wisdom blended with scientific knowledge and necessitates us to prepare for a paradigm shift in our life style and discard all those ways of living that lead to unsustainable approaches.

Sustainable agriculture, a holistic approach to farming system needs to be adopted embracing as many enterprises as possible where each one can complement the other in generating employment and augmenting income even in case of small and marginal farmers. We have to think of community approaches in input management and in protecting the natural resources and the crops and animals of the farmers with a collaborative and coordinated approach within a rural village.

This publication entitled *"Plant Sciences for Sustainable Agriculture and Rural Development"* is an outcome of the National Conference held earlier this year at the Balaghat Education Society's Arts, Science and Commerce College, Naldurg, Osmanabad (M.S.). It includes research papers presented at the Conference by researchers across the country. I am sure that the publication will be an excellent reference material useful to future researches.

The editors of this proceeding Dr. U. N. Bhale and Prof. V. S. Sawant are well known to me over the years. They are dedicated and committed workers in their field of study. I congratulate and complement them on this academic venture. I am sure that this proceeding will find a suitable place not only in libraries but will be highly useful to P.G. students, researchers, the teaching community and agriculturists in the country, at large.

B. F. Rodrigues

Professor
Department of Botany
Goa University, Goa

Contents

Chapter 1

Frontiers of Plant Sciences for Sustainable Agriculture and Rural Development

H.C. Lakshman[1]

Plants are vulnerable to environment stresses such as drought, cold, heat, salinity, etc., and unlike animals, they cannot take shelter to surround themselves with a potentially favorable growth environment due to their sessile nature, and they are left with two choices- adapt or die. As a result, plants are bestowed with two unique traits for survival. Firstly they have a highly organized defensive mechanism against the unfavorable climate and environmental conditions; secondly they have a highly effective propagation mechanism, which may be sexual or asexual reproduction.

Modern plant science has increased the productivity of agriculture. Farms can grow more on each acre and can do it more safely and fuel efficiently than ever before.

The use of agricultural technologies increases the productivity of land already cultivated. The overall number of species is declining at a historically high rate as the world's population increases and land is converted for industrial, domestic or agricultural use. Without the use of crop protection and biotechnologies (Brown, 2000).

India has inventoried over 47000 species of plants and 89000 species of animals over just 70 per cent of countries. The country is bestowed with immense agro-

---

1 P.G. Department of Studies in Botany, Microbiology Laboratory, Karnatak University, Dharwad – 580 003, Karnataka, India
E-mail: dr.hclakshman@gmail.com

biodiversity and a rich diversity in landraces/traditional cultivars/farmers varieties. A number of crop plants (384) are reported to be cultivated in India. A total of 49 indigenous major and minor crops have been reported in the History of Agriculture in India which include 5 cereals and minor millets, 4 pulses, 1 oilseed crop, 9 vegetables, 5 tuber crops, 11 fruits, 5 spices, 1 sugar yielding plants and 7 fiber crops. Amongst plants significant diversity has been recorded in pteridophytes with 1,022 species and Orchidaceae with 1,082 species.

Redesigning Irrigated Agriculture

Many civilizations flourished in river basin areas and produced unique cultural and scientific advances. On the whole, irrigation established a new foundation from which civilizations blossomed and profoundly shaped societal development (Kotharia, 2000). At the same time, history tells us that in a long, most irrigated-based societies fail. The inherent environmental instability of irrigated agriculture can weaken seemingly advanced cultures, rendering them less able to cope with political and social disturbances. Most of our irrigation bases are less than 50 years old, yet threats to the continued productivity have surfaced. One out of 5 ha of irrigation land is damaged by salt (Patil and Lakshman, 2003). In many areas problem of water

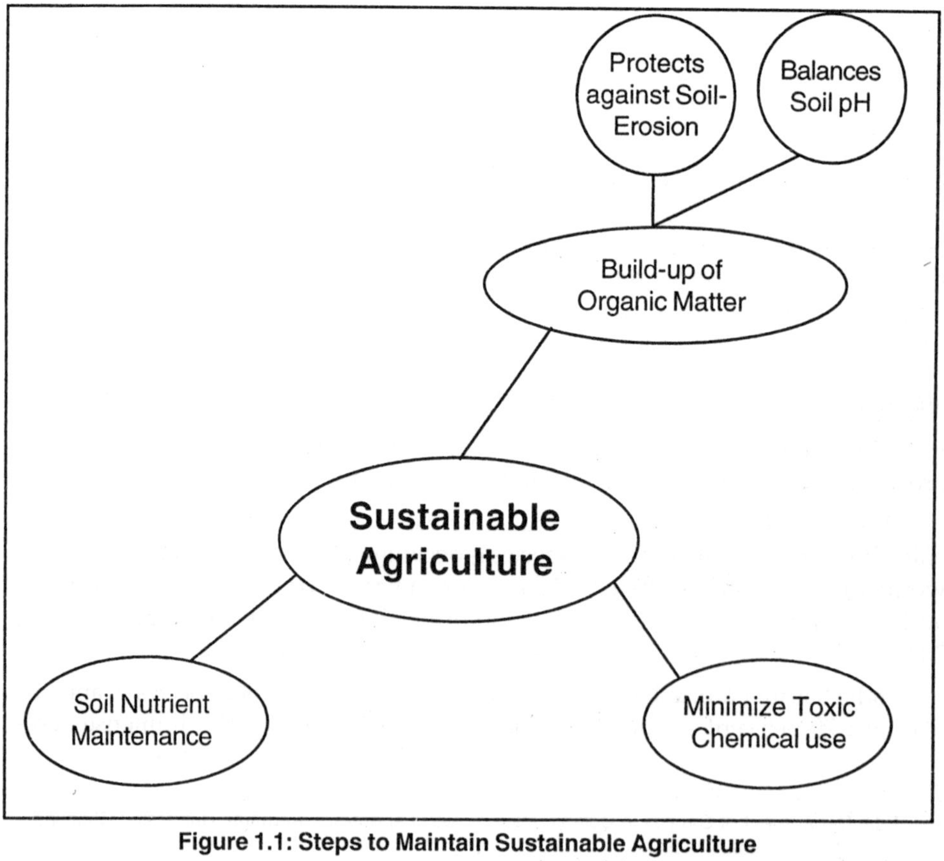

Figure 1.1: Steps to Maintain Sustainable Agriculture

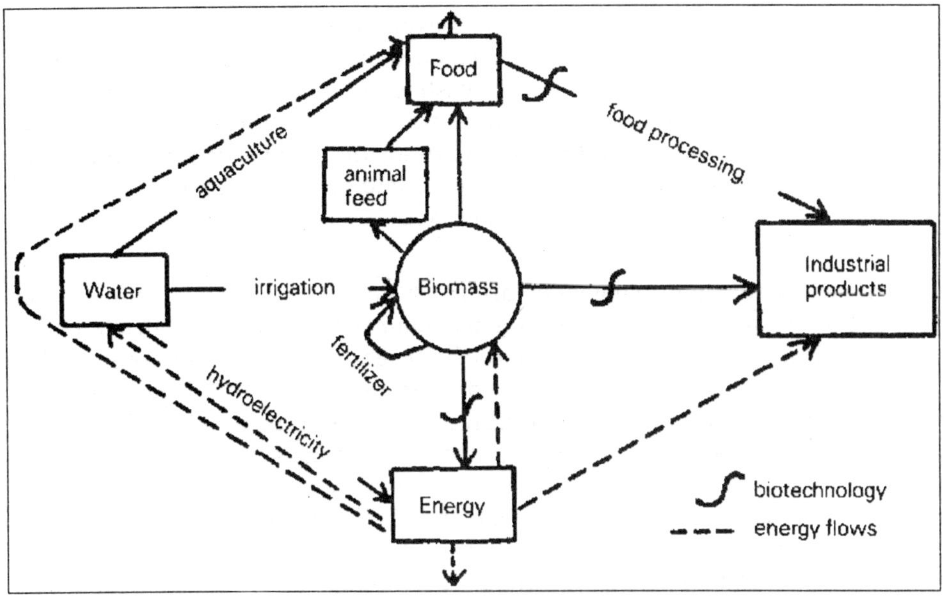

Figure 1.2: Components to Increase Biomass Production

logging has become a regular feature. More and more rivers are running dry for many months in a year leaving agriculture vulnerable to the reallocation of water to burgeoning cities and industries.

By now agriculture's principal challenge has been raising land productivity *i.e.* getting more crops out of each ha of land. But now challenge will shift to the boosting of water productivity *i.e.* getting more benefit from every liter of water devoted to crop production. We would have to save water loss due to evaporation and seepage in canals. There can be a long and growing list of measures that can increase agricultural water productivity which may include technical, managerial and institutional options, but more scientific inputs would have to be provident in evolving agronomic options such as selecting crop varieties with high yields per liter of transpired water, intercropping to maximize use of soil moisture, better matching crops to climate conditions and the quality of water available, sequencing crops to maximize output under conditions of soil and water salinity, selecting drought-tolerant crops where water is scare or unreliable, and above all breeding water-efficient crop varieties. Much attention would have to be given on diversification of crops in irrigated areas. We would also have to adopt new and economic methods of irrigation such as drip or sprinkle irrigation; therefore, research efforts should be diverted towards biosaline agriculture (ICAR, 2000). There is an urgent need to survey, conserve, and multiply new crops for saline habitats.

Restoration Ecology and Management of Natural Resources

Degradation of natural resources has been a global problem. Conversion of forest land into arable land or for other development activities such as urbanization and industrialization, intensive agriculture, over-exploitation, overgrazing, pollution of various kinds, mining and other anthropogenic activities have resulted in degradation of both the land and water resources. Harvesting of timber, collection of firewood, overgrazing, shifting, ignorance of proper soil conservation practices and crop rotation, non judicious use of fertilizers and pesticides, faulty irrigation and water management, discharge of industrial effluents and criminal disposal of sewage/ sludge are also responsible for soil and water resource degradation to a considerable extent.

The sustainable improvement of farming system is possible by halting further degradation of natural resource base and maintaining soil health through alternative agricultural practices such as adoption of efficient soil and water conservation measures, crop rotation, diversification of crops, reduction of tillage, integrated nutrient management involving bio-fertilizers as key constituents of beneficial micro organisms (Lakshman, *et al.*, 2005), integrated pest management, and efficient crop/ tree/livestock based systems. There should be fertility management strategies for specific problem soil *viz.* acid soils, saline and alkali soils, waterlogged soils, and hilly, arid and coastal soils. As per 1994 statistics, utilization of surface and groundwater in India is about 606 billion m^3, leaving 536 billion m^3 of utilizable water as unutilized for specific problem of the soil (Kumar, 1998).

The National Bureau of Plant Genetic Resources in New Delhi has responded to the increased world pressure for improvement in yield and content of drug plants by the pharmaceutical industry. One of the important contemporary developments for upgrading yield is the application of the genotype concept in medicinal plants, as has been done in agri-horticultural crops and intensified germplasm collections are being made.

Biofertilizer has been added from outside in the fields which is the most advanced biotechnology, necessary to support the developing organic agriculture, sustainable agriculture and non pollution agriculture. Biofertilizer is a natural organic fertilizer which provides all the nutrients required by the plants and helps to increase the quality of the soil with a natural microorganism environment and is eco- friendly, Biofertilizer contains a wide range of naturally chelated plant nutrients and trace elements, carbohydrates, amino acids and other growth promoting substances. It acts as a soil conditioner by stimulating microbial activity in the soil and improves or restores soil fertility (Lakshman, *et al.*, 2003). It improves air–water relationships in soil and makes soil less prone to compaction and erosion. It increases crop yield 20-30 per cent. It stimulates plant growth and it replaces chemical nitrogen and phosphorous by 25 per cent. Thus it reduces the cost towards fertilizers use, especially regarding nitrogen and phosphorous. It enhances biomass production 10-20 per cent.

The improvement of productivity of the major food, fiber and other crops has been impressive. The success is due to intensive research in plant breeding resulting in the development of new varieties with very high genetic yield potentials and suitable for wide range of agro-climatic conditions with the improvement of genetic engineering techniques. The biochemical and biotechnological processes have opened new frontiers to products of improved value from agricultural raw materials (Ott, 2003).

Perhaps the most revolutionary impact that comparative mapping brought to plant genomics was the establishment of systemic relationships among the genomes of the grass family (Graminae). The members of this family include the most economically important cereals and grasses such as rice (*Oryza sativa*), maize (*Zea mays*), barley (*Hordeum vulgare*), oat (*Avenā sativa*), bread wheat (*Triticum aestivum*), sorghum (*Sorghum bicolor*), foxtail millet (*Setaria italica*), and sugarcane (*Saccharum officinarum*). The initial cross genome comparisons with in this family, which aliened the genetic maps of rice, maize and wheat, revealed that the genome of rice, the smallest of the three species can be described in terms of genetic linkage blocks. Such data also the genome of rice, the smallest of the three species can be described in terms of genetic linkage blocks. Such data also confirmed the earlier hypothesis that maize is an ancient tetraploid as evidences by the occurrence of duplicated rice linkage blocks in its genome (Biswas, 2004). In rural development the plant science industry's efforts focus on protecting and improving yields, and preserving natural resources such as soil and water, using a wide range of technologies including chemical crop protection and biotechnology and novel techniques such as conservation agriculture.

Agricultural land is very much degraded due to over exploitation of the natural resources is commonly observed in rural area. The immediate effect of this is that the drinking water itself is the problem in the reason both in the hills and in the plains in both sides of the mountain though the rainfall is more than 1000 mm and up to 40,000 mm. Soil and moisture conservation, increasing knowledge of the ecosystem must be taught to rural folk. Efficient management of lands involving the local people creating and spreading rights and obligations in respect of natural resources use that involving NGO's and voluntary agencies must be strengthened. There is need of formation of agriculture committees(VAC's), meeting the requirement of fuel, fodder, minor forest produce and small timbers of the rural and tribal population, creating employment opportunity to the people living in and around the forest. There must be need of increased public participation, respect for the basic human rights, and improved popular access to education and information, and greater institutional accountability is essential elements for conservation.

Improvement of Existing and New Cultivated Plants

Biotechnology plays two roles in development of plant varieties. First the development of plant cell culture and gene analyses the period of conventional plant breeding by introducing plants with resistance to disease, pests and herbicides (Wate and Bodkhe, 2007). Diagnostic methods based on DNA hybridization or monoclonal antibodies for providing plant diseases in soil or the plant The improvement in quality of useful plant by increasing their nutritional value, tasty or chemistry example the in the proportion of certain essential amino acids like lysine in barley, the amount of fiber (as in tomatoes and flex) but also in the increase in amounts of chemically interesting compounds like oleic acids in sunflowers.

Green Revolution

Germplasm Exchange

More than 57,000 accessions in different agri-horti-silvicultural crops including multiplication trials, screening nurseries in wheat and rice were introduced from various countries

Table 1.1: Germplasm Exchange of Different Countries

Country	*Variety*
Philippines	blast resistant rice
Mangolia	cold tolerant wheat lines
U.S.A	Sorghum
Bangladesh	yellow sarson
U.S.A	Carthamus
U.K	Pea
Taiwan	Cowpea, Soybean
Malaysia	Ginger
Japan	Sweet potato

The following modern scientific methods could be more beneficial to enhance productivity these are given in steps as (Smith, 2009).

Concepts of Crop Improvement

1. Inter varietals or intra specific hybridization
2. Inter specific hybridization
3. Inter generic or distant hybridization

Methods for Crop Improvement

1. Plant tissue culture techniques.
2. Cell fusion method.
3. Somatic cell hybridization
4. Somoclonal variation
5. Organ culture techniques. Ex: cardamom (*Eletteria cardamomum*), pepper (*Piper nigrum*), ginger (*Zingiber officinalis*).

Plant Tissue Culture Techniques

1. Meristem culture
2. Embryo culture
3. Micropropagation
4. Cryopreservation- conservation and exchange of germplasm are hindered by perpetuation and spread of disease causing agents through planting

material. Meristem tip cultures have been used for cryopreservation of germplasm with 90 per cent tissue survival.

5. Embryo implantation
6. Ovule culture
7. Ovary culture
8. Intergeneric gene transfer

Table 1.2: Example of Agricultural Important Genes and Traits Transferred to Crop Plants by Interspecific or Intergeneric Hybridization

Crop Species	*Donor Species*	*Trait*
Cucubita pepo	*C. lundelliana*	Mildew resistance
Gossypium hirsutum	*G. tomentosum*	Nectorless (decreased incidence of bolrot)
Lycopersicon esculentum	*L. birsutum*	Bacterial canker resistance
Oryza sativa (rice)	*O. nivora*	Grassy stunt virus resistance
Santalum tuberosum	*S. acaule*	Potato virus resistance
Triticum aestivum	*Aegilops comosa*	Strip rust resistance
Zea mays (maize)	*Tripsacum* spp.	Northern corn leaf blight resistance

Agriculture sector today, needs to achieve an increase in the yield per hector in sustainable manner at an affordable cost. When the world faced similar challenges at the beginning of the 20th century, Biotechnology has come up as a remedy and is expected to lead the way to achieve increase in food production in the 21th century (Kameshwara Rao, 2011).

The primary function of National Bureau of Plants and genetic Resources (NBPGR) has been to widen the genetic base of economic plants and to make available diverse germplasm top breeders for crop improvement work. ICAR crop-based institutes, crop-coordinated projects and scientists in agricultural universities and state department of agriculture.

Rural Development

During the first quarter of the present century, it was realized that if the poverty of India's teeming millions is to be removed, more attention had to be given to rural areas where 80 per cent of Indian population is living. Therefore several experiments in rural re-construction progressive work was undertaken.

Integrated Rural Development Programme (IRDP)

The Integrated rural development focused on the target group comprising small and marginal framers, agricultural labourers and rural artisans, where economic improvement was an important concern of rural development. Farmers should be encouraged to attend Krishi Mela in different agricultural universities and training programme should be given on poultry, and the use of organic farming (Palaniappan and Annadurai, 2003). They should be encouraged to give good education to girl child.

Prawn Culture

The second product of aquaculture in India is prawn culture. The prawn fishery of India is next to that of U.S.A. in the world. Prawn is commercially known as shrimp. The prawn is belonging to the families Penaeidae, Pandalidae, Hippolytidae, Sergaetidae and Palaemonidae. Some species are good for culture while several others are caught in very small quantities, but they also add to the total catches.

Pearl Culture

Although a number of bivalves (molluscs) have ability to produce pearl under suitable climatic conditions but high quality of pearls are obtained from pearl oysters of genus *Pinctada*. A number species of this genus *P. vulgaris, P. margaritifera, P. anomioibes* and *P. atropurpurea*.

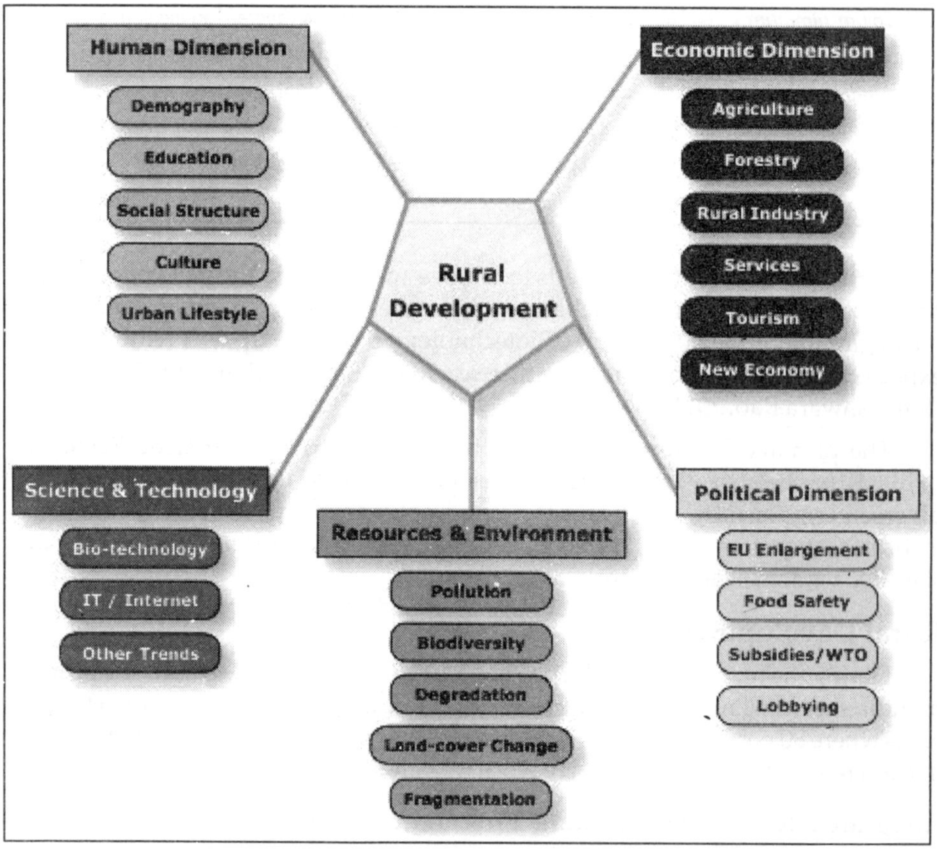

The I.R.D.P aimed at integrating field programmes reflecting the economic activities of the rural families whose employment and development were the basic objective. Programmes of agriculture, animal husbandry, fisheries and forestry development were brought into focus by maximizing land, water and cropping facilities in a holding through a mixed farming approach. This was the core of the

I.R.D. Programme for implementing I.R.D. Programme at the district level. District rural development agencies have been set up in all the districts in the country.

Intensive Agricultural Area Programme (IAAP)

The Agricultural Production Board in the Union Ministry of Food and Agriculture had agreed that about 20-25 per cent of cultivated area of the country should be selected for an intensive agricultural development. The object of Intensive Agricultural Area Programme was to increase the production of main crops in selected areas by an intensive and co- ordinate use of various aids to production.

Marginal Farmers and Agricultural Labours Agency (MFALA)

Marginal Farmers and Agricultural Labours Agencies were set up on recommendation made by the Rural Credit Review Committee (1969). The principle objective of the M.F.A.L.A programme was to assist the marginal cultivators in making the maximum productive use of their small holdings by undertaking animal husbandry, dairy and horticulture etc.

Small Farmers Development Agency (SFDA)

The establishment of Small Farmers Development Agencies was an attempt to help small farmers in selected areas; the genesis of these agencies lies in the report of All India Rural Credit Review Committee.

The main functions of S.F.D.A are:

1. To identify small farmers.
2. To supply various inputs required by the small farmers.
3. To help in credit facilities
4. To help farmers for transportation, storage, processing and marketing of their produce.
5. Arrangement of irrigation facilities for small farmers.

Drought Prone Area Programme (DPAP)

Drought Prone Area Programme was initiated during the fourth land period in 1970-71. The drought prone areas were selected on the basis of irrigated areas, low or erratic distribution of rain fall and high periodicity of drought (Powlson, *et al.*, 1998).

The main functions of D.P.A.P are

1. To utilize all resources in the affected areas.
2. To improve the standard of living of rural drought affected people.

Mechanical Measures

The mechanical measures used for controlling soil erosion are to construct terraces and to check gully erosion, check dams are constructed.

Crop Rotation

Crop rotation helps us to check weeds, pests and diseases. Different crop plants have different types of root systems, so that successive crops tap nutrients from different depths of the soil. Crops have their specific weed and thus rotation brings about a natural reduction in the number of weeds. The requirements of the successive crops differ and hence by rotation the soil is improved.

The main plan was to restore a proper ecological balance in drought prone areas through:

1. Development and management of irrigation resources.
2. Soil and moisture conservation and afforestation programme.
3. Re-structuring of cropping pattern and pasture development.
4. Liverstock development.
5. Development of small and marginal farmers.

Food for Work Programme

The basic objective of the programme were to generate additional employment in rural areas through utilization of available stock of food grains and to create their durable community assets which could strengthen the infrastructure in rural areas for socio- economic development (Swaminathan, 2005). Besides being successful in achieving our objectives, the programme resulted in many other benefits to the rural poor like price stabilization of food grains, ensuring minimum wages to the workers and providing work through out the year.

Jawahar Rozgar Yojna

Unemployment and poverty are the basic problems before us. About 75 per cent of our population lives in rural areas, therefore the intensity of the problems is more in rural areas. The population of schedule caste and schedule tribes is poorest among the poor. Again the problem of unemployment in these sections is very much. To tackle these problems, specific programmes, are the needed.

Eighty per cent of total finances for the Yojna will be defrayed by the central government and rest twenty per cent will borne by the state government. The important aspect of this scheme is that the village panchayats will be pivot around which the scheme will revolve. As poverty is the product of economic, social, political, cultural and environmental factors, a package of programmes is required to mitigate the poverty.

There are following benefits by crop insurance to farmers.

1. There is big protection to farmers against the failure of crops and stabilize their incomes.
2. Crop insurance improves their credit worthiness in securing loans from a credit agency.
3. Crop insurance provides greater confidence to farmers to venture upon the adoption of modern technology involving larger expenditure on modern inputs and greater risk due to natural hazards.

Certain problems arise in carrying out the crop insurance scheme to the desired extent.

1. It is administratively difficult to inspect crops and determine the extent of loss.
2. It is expensive to determine the rate of premium for individual farmers.
3. Chances of corruption among the lower strata of the revenue and insurance agents cannot be ruled out.

References

Biswas, Anirban (2004). Productivity in Indian agriculturae. *TIC Services*, pp. 1–2.

Brown, L.R. (2000). Challenges of the century. In: *State of the World 2000*. The World Watch Institute W.W. Norton and Co., New York, pp. 3–21.

Indian Council of Agricultural Research, 2000). Agric. Statistics Division, Ministry of Agric., Govt. India, New Delhi.

Kameswara Rao, C. (2011). Modern biotechnology, biodiversity and sustainable development. In: *Souvenir of Biodiversity and Biotechnology for Sustainable Development*, (Ed.) H.C. Lakshman, pp. 138.

Kothari, D.S. (2010). Science and Universities. *Everyman's Science*, 155(3): 145–157.

Kumar, P. (1998). Food demand and supply position for India. *Agric. Econ. Policy Paper 98–01*, IARI, New Delhi.

Lakshman, H.C., Patil, G.B. and Hosamani, P.A. (2003). The effect of VA mycorrhizae and different levels of nitrogen on growth and biomass production of black pepper. *Spice India*, 16(7): 26–29.

Lakshman, H.C., Hosamani, P.A. and Kadam, L.B. (2005). Microorganisms and their role in phosphate solubilization. In: *Biotechnological Applications in Environmental Management*. B.S. Publications, Hyderabad, pp. 71–79.

Ott, K. (2003). The case for strong sustainability. In: *Greifswald's Environmental Ethics*, (Eds.) K. Dott and P. Thapa. Greifswald: Steinbecker Verlag, Ultrich Rose.

Palaniappan, S.P. and Annadurai, K. (2003). *Organic Farming: Theory and Practice*. Scientific Publisher (India), Jodhpur, pp. 10–22.

Patil, G.B. and Lakshman, H.C. (2003). Effect of AMF and saline water with and without additional phosphate on elusine coracana (Finger millet. *Indian J. Env. and Ecopl.*, 7(3): 477–482.

Powlson, D.S., Poulton, P.R. and Gaunt, J.L. (1998). The role of long-term experiments III agricultural development. In: *Proc. of National Workshop on Long-term Soil Fertility Management through Integrated Plant Nutrient Supply*, (Eds.) A. Swamp *et al*. Indian Institute of Soil Science, Bhopal, India, p. 1–15.

Swaminathan, M.S. (2005). 2006–Year of Agricultural Renewal. *The Hindu*, 31st December.

Smith, J.E. (2009). *Biotechnology*, Fifth Edition. Cambridge University Press, New York, 268 pp.

Wate, S.R. and Bodkhe, S. (2007). Biotechnology in environment management. In: *Sustainable Environmental Management*, (Eds.) L.V. Gangawane and V.C. Khilare. Daya Publishing House, New Delhi.

Chapter 2

In vitro Culture of Mycorrhizal Fungi (AM) Fungi: An Overview

***K.P. Radhika*[1] *and B.F. Rodrigues*[1*]**

Introduction

The technique of *in vitro* cultivation of root organs has developed over the past few decades, opening up new ways of studying plant-fungal associations. Despite the large array of applications, the use of excised root organs to study the AM symbiosis has major limitations *viz.*, absence of photosynthetic tissues, abnormal hormonal balance and modified physiological source-sink relationships (Fortin *et al.*, 2002). The first *in vitro* culture associating an AM fungus with a whole plant dates back to early sixties (Mosse, 1962). In the following two decades several authors reported *in vitro* association of AM fungus with host plant in liquid and agar media (Mosse and Philips, 1971; Hepper, 1981; MacDonald, 1981; Strullu and Romand, 1986) with improvement in technologies which allowed the successful cultivation of a few AM fungal strains. The present paper reviews the developments in *in vitro* culturing of AM fungi.

Monoxenic Culture of AM Fungi

Monoxenic culture of AM fungi has markedly improved understanding of their symbiosis. In the past 15 years, increasing literature has been devoted to this intimate

---

1 Department of Botany, Goa University, Taleigao Plateau, Goa – 403 206, India
*E-mail: felinov@gmail.com

plant fungal association using various AM fungi *in vitro* cultivation systems with different hosts, AM fungal propagules and growth media. Using this methodology, the traceability of monoxenic cultures (root and fungi) can be followed for several generations by transferring a mature section of a colony to a new substrate. Maintained in Petri plates, these cultures can provide continuous access to contaminant-free fungal material (Fortin *et al.*, 2002).

Root Organ Cultures (ROC) were originally developed by White (1934) by obtaining continuous culture of non-transformed tomato (*Solanum lycopersicum* Mill.) roots in liquid media. Mosse and Hepper (1975) established the first *in vitro* culture (on solid media) between ROC and AM fungi using excised non-transformed roots of *Solanum lycopersicum* and *Trifolium pratense* where the roots were colonized and the fungal life cycle completed by the production of spores. Further, the variations in the culture media by Strullu and Romand (1986) and Becard and Fortin (1988) (M medium), Chabot *et al.* (1992b) (MW medium) and St-Arnaud *et al.* (1996) (MC medium) has allowed both transformed and non-transformed ROCs to get mycorrhizal *in vitro* and subsequently maintain them in continuous culture.

Research using monoxenic cultures in late 1980s began mainly with *Gigaspora margarita* DAOM 194757 (Becard and Fortin, 1988). The long term behaviour of *Gi. margarita* on Ri T-DNA transformed carrot roots indicated that 80 per cent of fungal colonization units were produced during root aging period. Mugnier and Mosse (1987) and Becard and Fortin (1988) used Ri T-DNA transformed carrot root as host in ROC system. ROC of Ri T-DNA transformed carrots roots with AM fungi have been widely used to study mycelium architecture, dynamics of fungal development and spore ontogeny (Becard and Fortin, 1988).

In 1990s AM fungal isolate *Glomus intraradices* DAOM 197198 was mostly used as this AM fungal isolate was well adapted to monoxenic culture conditions, producing significant amounts of tissue and easy to replicate (Chabot *et al.*, 1992b). The use of sun bags *in vitro* system is an alternative to obtain large scale AM inoculum without any contamination (Diop, 1993). Tiwari and Adholeya (2003) demonstrated large scale production of AM spores by culturing root organs and AM fungi in small containers. Large scale cultivation of AM fungi has also been performed in an aircraft bioreactor (Jolicoeur *et al.*, 1994), in a mist bioreactor with perlite as substrate (Jolicoeur, 1999) and in a bioreactor containing solid gelled medium (Fortin *et al.*, 1996).

Kandula *et al.* (2006) reported first monoxenic culture of *Scutellospora calospora* with Ri-TDNA transformed carrot (*Daucus carota* L.) roots. Although spores were not produced in abundance, a standard method for dual culture of *S. calospora* and completion of the life cycle was established. The use of monoxenic cultivation of *S. calospora* provided a unique tool to understand the critical colonization events for spore formation. Successful root organ culture of *S. reticulata* has shown that it produces mycelium of two architectural patterns, one related to root colonization and the other to resource exploitation (De Souza and Declerck, 2003).

St-Arnaud *et al.* (1996) developed compartmentalization system by separating mycelium from excised mycorrhizal root on the other side. This improvement allowed the study of lipid metabolism (Bago *et al.*, 2002), transport of element to roots (Nielsen *et al.*, 2002) and to isolate microbial free AM fungal mycelium and spores for molecular analysis (Pawloska and Taylor, 2004). Douds *et al.* (2000) studied the carbon uptake and transport of AM fungi using split plate method *in vitro* using Ri T-DNA transformed roots and observed fundamental differences in carbon metabolism of different forms of hyphae of *G. intraradices*. However the transformed roots offer an advantage to present a greater growth potential than non transformed roots in culture media with limited supply of nutrients (Fortin *et al.*, 2002).

Some attempt had been made to study AM interactions with AM monoxenics. Negative competition was reported in a co-culture of *Gi. margarita* and *Gi. gigantea* (Douds and Becard, 1993), while no antagonism was reported in a co-culture of *G. intraradices* and *Gi. margarita* (Tiwari and Adholeya, 2002). Co-culture technique offers an opportunity for an insight into various phenomena like interaction, competition and dominance occurring among different AM fungal genera. This allows mass production of multiple AM fungal isolates under single root organ culture.

Gadkar *et al.* (2006) developed a container with Petri plate containing ROC to initiate fungal proliferation in a separate compartment filled with sterile expanded clay balls for large scale production of extra-matrical soluble compounds. Nogales *et al.* (2010) carried out *in vitro* interaction studies between *G. intraradices* and *Armillaria mellea in vitro* micropropagated plantlets of grape vine and observed that AM symbiosis has a protective effect at plant level, reducing disease symptoms. Cano *et al.* (2008) successfully obtained freshly colonized excised roots *in vitro* using bi-compartmental Petri plates with a mycorrhizal donor organ culture developing in a root compartment.

Monoxenically grown AM fungi allow non destructive observation of fungal colonies, mycelium organization and sporulation steps. With this technology, original data on mycelium development (St-Arnaud *et al.*, 1996), sporulation dynamics (Declerck *et al.*, 2001) and spore ontogeny (Pawloska *et al.*, 1999) have improved our understanding of AM fungal propagation process and life cycles.

Selection of Host Roots for *in vitro* Studies

Irrespective of the type of root system chosen, success in establishing a mycorrhizal culture depends on the physiological state of the host root. Large scale production of *G. intraradices* spores *in vitro* was first attempted on ROC (Douds, 2002) and later extended to plant systems. Initially carrot roots were used as host but in recent years chicory (*Cichorium intybus* L.) and barrel medic (*Medicago trunculata* Gaertum) have been successfully established to culture AM fungi (Fontaine *et al.*, 2004). Tiwari and Adholeya (2003) reported that a change in root clone could impact AM fungal spore production. Other hosts such as banana (Koffi *et al.*, 2009), were also found effective for large scale production of spores.

Recently, however with the use of non-transformed roots, availability of plant root mutants and development of new methods for culturing whole plants *in vitro* has increased opportunities for future research concerning *in vitro* culture of AM fungi.

In whole plant *in vitro* system either the roots or the entire plant itself is grown under aseptic conditions (Voets *et al.*, 2005). A derived plant *in vitro* production system has recently been detailed in a patent proposal (Declerck *et al.*, 2009) where the pre-inoculated *in vitro* produced plant (Voet *et al.*, 2009) is individually introduced into a sterile growth tube and nutrient solution circulates in a closed system flowing into mycorrhizal roots. This new method overcomes the controversy that has previously arisen due to the use of using only roots as hosts (Bago and Cano, 2005). Selection of host root is also influenced by the type of morphological development of AM fungus on colonizing the plant root (Arum and Paris type), and its effect on plant fungal metabolism and gene expression (Cano *et al.*, 2005).

Inoculum Types for *in vitro* Studies

Isolated spores, vesicles, mycelia and mycorrhizal roots are able to initiate AM symbiosis but the preferred inoculum includes chylamdospores of *Glomus* sp. and monosporic azygospores of *Gi. margarita* due to their ability to recover, sterilize and germinate the propagules (Diop, 2003). Intra-radical forms of AM fungi like intra-vesicles have higher inoculum potential than other AM fungal propagules (Moses, 1988). In monoxenic culture, proper selection and efficiency of the sterilization process is the key to success.

Surface Sterilization of Spores and Mycorrhizal Root Fragments

Spores are usually collected from the field, or from pot cultures, by wet sieving and chosen individually under a dissecting microscope using a micropipette. Before being used as *in vitro* inoculum, spores are surface sterilized (Bécard and Piche, 1992) to eliminate all contaminants. A solution containing a strong oxidizing agent, chloramine T and a surfactant (tween 20) is widely used to sterilize AM fungal spores. To overcome dormancy, spores are stored at 4°C, either in distilled water, on water agar or on 0.1 per cent $MgSO_4{\cdot}7H_2O$ solidified with 0.4 per cent gellan gum. If spores fail to germinate within 20 days it is concluded that either the sterilization treatment is too strong or the spores are immature, dormant or dead. Spores of some AM fungal species require cold stratification (4°C) prior to germination (Smith and Read, 1997).

Mycorrhizal roots are used to initiate monoxenic cultures because of their high susceptibility to colonization. Young and healthy roots should be chosen for *in vitro* culture establishment, as it is easy to locate presence of vesicles in such roots. The roots are then disinfected in an ultrasonic processor under a laminar-flow hood (Declerck *et al.*, 1998). This method has been successfully used to isolate numerous AM fungal species (Declerck *et al.*, 1996; Dalpé 2001). Treatment duration and reagent concentrations are adapted to specific situations *viz.*, host plant, root age and contamination level.

Culture Media

The M medium (Becard and Fortin, 1988) is a modified White's medium initially developed for tomato root-organ cultures (Butcher, 1980). The macro element composition of White's medium is considerably lower than that of MS and B5 media, commonly used for *in vitro* plant cultures (Bécard and Piché, 1992). However, this dilute medium is adequate for root growth. The MSR medium is a modified medium which was developed to optimize the growth of the intraradical phase of the fungus *in vitro* (Declerck *et al.*, 1998). Almost 30 AM fungal isolates from the Acaulosporaceae, Gigasporaceae and Glomaceae are now successfully grown on these media.

Several studies have concentrated on the effect of medium pH on the establishment of symbiosis. *Glomus mosseae* showed germination with above pH 7 (Douds, 1997) whereas in other studies germination of *G. mosseae* (Mugnier and Mosse, 1987) and *G. intraradices* (Mosse, 1988) was inhibited at low medium pH and hyphae grew after pH was increased.

In vitro Growth

In monoxenic culture, the pre-symbiotic growth is characterized by the formation of runner hyphae. After a few weeks, in the absence of host root, the germination of AM propagules ceases with the formation of septate hyphae and retracted cytoplasm. This could be due to nutrition (Hepper, 1983), chemical treatments (Gianiazzi-Pearson *et al.*, 1989) and genetic factors. Low amount of natural and synthesized flavanoids, besides certain amino acids are known to positively stimulate fungal growth (Hepper and Jacobsen, 1983). Volatile compounds like CO_2 and root exudate promote hyphal elongation (Carr *et al.*, 1985). Hildebrant *et al.* (2002) found the presence of slime forming bacteria *Paenibacillus validus* on surface sterilized spores of *G. intraradices*. These bacteria stimulate the growth of *G. intraradices* until spore formation in the absence of plant tissue. Bruce *et al.* (2000) isolated a semi-purified fraction called "Branched Factor" from mycotrophic plant species that are found to be active in stimulation of germination and nuclear division of *Gigaspora* species.

Dual root organ culture grown under *in vitro* conditions have been used for studying AM extra-radical mycelium (Bago *et al.*, 1996). Although monoxenic AM cultures are artificial, they provide a better understanding of AM fungal biology and behaviour. Recently, it has proved suitable for the physiological study of AM extra-radical mycelium and biology. AM monoxenic culture studies are focused mainly on extra-radical mycelium biology, but recent reports also indicate its role in studying intra-radical colonization and fungal host interactions (Declerck *et al.*, 2000).

Methodologies for *in vitro* Cultivation of AM Fungi

The process of culturing and maintaining monoxenic cultures of AM fungi is divided into four main steps *viz.*, selection of adequate AM fungal propagules; sampling, disinfection and incubation of the propagules on a suitable growth medium; association of the propagules with a suitable host root; and sub cultivation of the AM fungi. Prior to these steps are selection of an appropriate culture system, the preparation of synthetic culture media and the management of the host root (Cranenbrouck *et al.*, 2005) (Figure 2.1).

Fundamental and Practical Studies

Until now root segments and spores isolated from open pot cultures (Glimore, 1968) have been the source of AM inoculum for research purposes. This type of inoculum requires large space and is prone to contamination. Hence production of

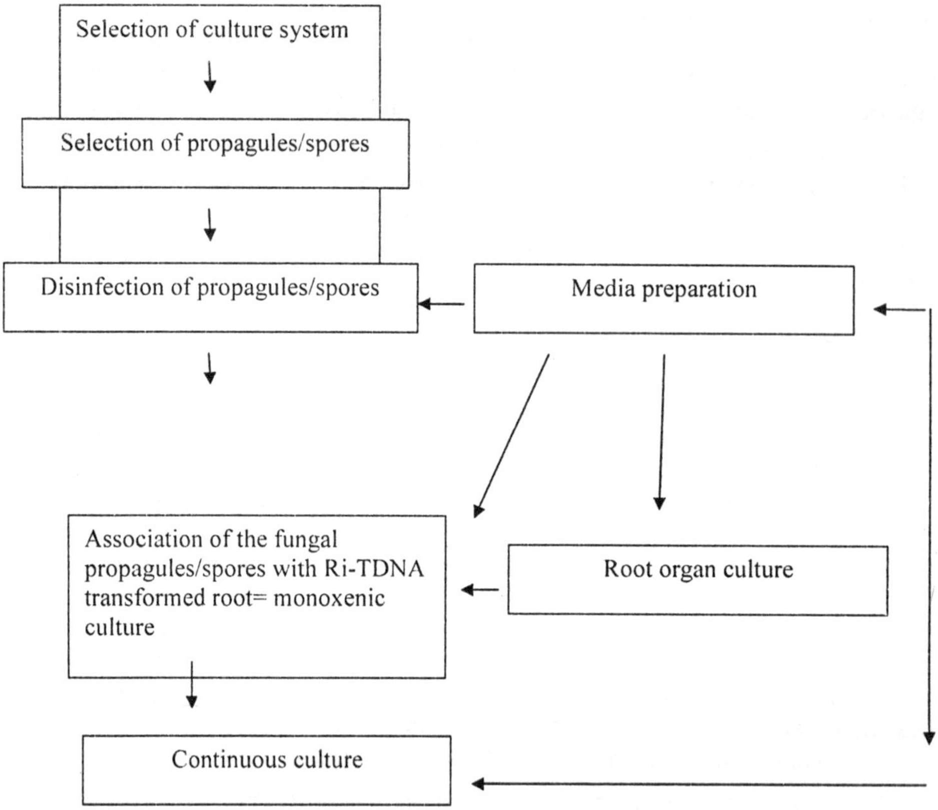

Figure 2.1: *In vitro* Culturing of AM Fungi (Cranenbrouck *et al.*, 2005)

spores under aseptic conditions is the most promising way of obtaining high quality pathogen free inoculum for research purposes.

Although *in vitro* culture is an artificial system, it is a valuable tool for the study of fundamental and practical aspects of AM symbiosis, complementing other experimental approaches. Development of extra-radical mycelium under aseptic conditions is preceded by the formation of arbuscule-like structures (ALS) (Bago *et al.*, 1998a) or branched absorbing structures (BAS) (Bago *et al.*, 1998b). These are finely branched hyphae and play an important role in nutrient uptake by increasing nutrient supply to the developing spore. Diop (1995) established a germplasm bank of AM monoxenic culture in association with isolated tomato or transformed carrot roots. The propagules (spores, hyphae, colonized roots) produced germinate and re-colonize new plants efficiently. Encapsulation stabilizes the biological properties of mycorrhizal roots and isolated vesicles or spores (Strullu *et al.*, 1991; Declerk *et al.*, 1996b). This immobilization also preserves the colonization ability of AM propagules in both *in vitro* and *in vivo* assays.

The use of AM root-organ culture as a simplified model system has been particularly useful for the investigation of at least two physiological aspects, signaling between the symbiotic partners and metabolism of the fungus. This has been possible because most parameters *viz.*, host root, fungal inoculum, and the physical, chemical and microbiological environment are strictly controlled. Under these conditions, detailed non-destructive observations at the morphological and cellular levels are made possible. The AM root-organ culture system also allows the production of pure fungal biomass, at various symbiotic stages and in sufficient quantities, for further cytological or biochemical analyses (Fortin *et al.*, 2002).

The increasing number of species of AM fungi cultivated *in vitro*, and the possibility of continuous cultivation and cryopreservation, has led to the development of an international collection of *in vitro* AM fungi–the Glomales *in vitro* collection (GINCO). This collection has resulted from collaboration between the Mycothèque de l'Universite C holique de Louvain (MUCL, Belgium) and the Eastern Cereal and Oilseed Research Centre (ECORC, Agriculture and Agri-Food Canada) which is responsible for the Canadian Collection of Fungal Cultures (CCFC/DAOM, Canada). GINCO aims to conserve biodiversity and provide high-quality, contaminant-free AM fungal inocula for scientific research. GINCO, in collaboration with a team of scientists working on AM fungal physiology, biochemistry, taxonomy and ecology intends to increase the number of taxa available by offering specialized training, and also by developing an international network of collaborative research (Fortin *et al.*, 2002).

Continuous *in vitro* Culture to Obtain Pure Mycorrhizal Cultures

The first continuous culture was achieved by Strullu and Romand (1986) and is now commonly used for a wide range of *Glomus* (Strullu *et al.*, 1997; Declerck *et al.*, 1998) and *Gigaspora* species (Bécard, unpublished data). Continuous cultures are obtained by transferring mycorrhizal roots to fresh medium either with spores (St-Arnaud *et al.*, 1996) or without (Declerck *et al.*, 1996). Following this transfer, the pre-existing root–fungus association continues to proliferate. Older mycorrhizal roots are transferred to a Petri plate containing an actively growing root (Strullu *et al.*, 1997). A major challenge in monoxenic culture is to maintain this culture over generations *i.e.* to obtain continuous monoxenic culture. This is a crucial issue which would ensure a) correct functioning of *in vitro* symbiotic interaction, and b) allow clonal fungal material essential for molecular studies to be obtained while minimizing undesirable external variation.

The best example of successful continuous AM monoxenic culture is *G. intraradices* DAOM 197198, which has been maintained continuously since 1992. A major step which pushed mycorrhizal research forward was the two compartment monoxenic system where extra radical root free fungal mycelium is obtained. Knowledge on metabolism of AM fungi has been greatly benefited by the use of compartmentalized monoxenic cultures as there is a possibility to precisely control the media composition for extra radical mycelium growth (St-Arnaud *et al.*, 1996).

An increasing number of AM fungal species have been cultured monoxenically as described in a number of papers reporting the production of root-organ cultures. These include *Acaulospora rehmii* (Dalpé and Declerck, 2002), *Gigaspora rosea* (Bago *et al.*, 1998), *Gigaspora margarita* (Miller-Wideman and Watrud, 1984), *Gigaspora gigantea*

(Gadkar *et al.*, 1997), *Glomus etunicatum* (Schreiner and Koide, 1993), *Glomus intraradices* (Chabot *et al.*, 1992; St-Arnaud *et al.*, 1996), *Glomus versiforme* (Declerck *et al.*, 1996; Diop *et al.*, 1994), *Glomus caledonium* (Karandashov *et al.*, 2000), *Glomus fasciculatum* and *Glomus macrocarpum* (Declerck *et al.*, 1998), *Glomus proliferum* (Declerck *et al.*, 2000), *Glomus deserticola* (Mathur and Vyas, 1995) and *Glomus fistulosum* (Gryndler *et al.*, 1998).

Cryopreservation of *in vitro* Produced Spores

The continuous culture method allows AM fungi to be maintained monoxenically over long periods and serves to produce and store AM fungal germplasm. However, AM fungi in pot culture method do not ensure genetic stability of the fungal material over generations (Declerck *et al.*, 1998). Several methods of long term storage have been developed to overcome this problem, one of which consists of storing cultures at 4°C. Although information exists concerning the survival of pot-culture inoculum (Douds and Schenck 1990), there are little or no data concerning monoxenically produced inocula. At 4°C, fungal and root metabolism is slowed, but not halted. Addy *et al.* (1998) were the first to demonstrate that monoxenically grown extraradical mycelium of *G. intraradices* was able to survive exposure to 12°C. These authors slowly cooled cultures before freezing. Recently, Declerck and Angelo-van Coppenolle (2000) developed a cryopreservation technique based on the entrapment of monoxenically produced spores of *G. intraradices* in alginate beads.

Conclusion

In vitro culture systems offer versatility and potential application in many fields of AM research. The availability of abundant root free fungal mycelium and spores has proven to be useful in morphological, biochemical, physiological and molecular studies (Fortin *et al.*, 2002). Increased knowledge on mycorrhizal symbiosis has been obtained during recent years with the use of monoxenic cultures. *In vitro* grown AM fungi constitute a reliable material for species characterization and description and for evolutionary and interspecific studies (Declerck *et al.*, 2001).

References

Addy, H.D., Boswell, E.P. and Koide, R.T. (1998). Low temperature acclimation and freezing resistance of extraradical VA mycorrhizal hyphae. *Mycol. Res.* **102**: 582–586.

Bago B., Pfeffer P.E., Zipfel W., Lammers P. and Shachar-Hill, Y. (2002). Tracking metabolism and imaging transport in arbuscular mycorrhizal fungi. *Plant Soil.* **244**: 189–197.

Bago, B. and Cano, C. (2005). Breaking myths on arbuscular mycorrhizas *in vitro* biology. In: *In vitro culture of mycorrhizas*. (Eds) Declerck, S., Strullu, D.G., Fortin, J.A. Springer-Verlag, Heidelberg, pp. 111-138.

Bago, B., Azcón-Aguilar, C. and Piche, Y. (1998a). Extraradical mycelium of arbuscular mycorrhizae: the concealed extension of roots. In *Radical biology: advance and perspectives on the function of plant roots*. (Eds) H.E. Flores, Lynch J.P., Eissenstat,

D. Current Topics in Plant Physiology, American Society of Plant Physiologists, Rockville, Md. Series 18, pp. 502–505.

Bago, B., Azcon-Aguilar, C. and Piche, Y. (1998). Architecture and developmental dynamics of the external mycelium of the arbuscular mycorrhizal fungus *Glomus intraradices* grown under monoxenic conditions. *Mycologia* **90**: 52-62.

Bago, B., Azcon-Aguillar, C., Goulet, A. and Piche, Y. (1998b). Branched absorbing structures (BAS): a feature of the extraradical mycelium of symbiotic arbuscular mycorrhizal fungi. *New Phytol.* **139**: 375–388.

Bago, B., Vierheilig, H., Piche Y. and Azcon-Aguilar, C. (1996). Nitrate depletion and pH changes induced by the extraradical mycelium of the arbuscular mycorrhizal fungus *Glomus intraradices* grown in monoxenic culture. *New Phytol.* **133**: 273-280.

Becard, G. and Piche, Y. (1992). Establishment of vesicular-arbuscular mycorrhiza in root organ culture: review and proposed methodology. In: *Techniques for the Study of Mycorrhiza.* (Eds.) Norris, J., Read, D., Varma, A. Academic Press, New York. pp. 89-108.

Becard, G. and Fortin, J.A. (1988). Early events of vesicular–arbuscular mycorrhiza formation on Ri T-DNA transformed roots. *New Phytol.* **108**: 211–218.

Butcher, D. N. (1980). The culture of isolated roots. In: *Tissue Culture Methods for Plant Pathologists.* Ingram, D.S., Helgelson, J.P. (Eds). Blackwell Scientific, Oxford. pp. 13–17.

Buee, M., Rossignol, M., Jauneau, A., Ranjeva, R. and Becard, G. (2000). The pre-symbiotic growth of arbuscular mycorrhizal fungi is induced by a branching factor partially purified from plant root exudates. *Mol. Plant Microbe. Interact.* **13**: 693–698.

Caar, G.R., Hinkley, F., Le Tacon, F., Hepper, C. M., Jones, M. G. K. and Thomas, E. (1985). Improved hyphal growth of two species of vesicular arbuscular mycorrhizal fungi in the presence of suspension-cultured plants cells. *New Phytol.* **101**: 417-426.

Cano. C., Dickson, S., Gonzalez-Guerrero, M. and Bago, A. (2008). *In vitro* cultures open new prospects for basic research in arbuscular mycorrhizas. In: *Mycorrhiza.* Varma, A. (Ed). Springer-Verlag, Berlin, pp: 627–654.

Cano, C., Dickson, S., Guerrero, G.M. and Bago, A. (2005). *In vitro* culture open new prospects for basic research in Arbuscular mycorrhizal research. In: *Mycorrhiza State of Art, Genetics, Molecular biology, Ecofunction Biotechnology, Eco-physiology, structure and Systematics.* Verma A. (Eds 3rd). Springer-Verlag, Heidelberg. pp: 627-654.

Chabot, S., Becard, G. and Piche, Y. (1992). Life cycle of *Glomus intraradix* in root-organ culture. *Mycologia.* **84**: 315–321.

Chabot, S., Bel-Rhid, R., Chenevert, R. and Piche, Y. (1992b). Hyphal growth promotion *in vitro* of the VA mycorrhizal fungus *Gigaspora margarita* Becker and Hall, by the activity of structurally specific flavonoid compounds under CO_2 enriched conditions. *New Phytol.* **122**: 461–467.

Cranenbrouck, S., Voets, L., Bivort, C., Renard, L., Strullu, D.G. and Declerck, S. (2005). Methodologies for successful aseptic growth of arbuscular mycorhizal fungi on root-organ. In: *In vitro culture of mycorrhizas.* Declerck, S., Strullu, D.G., Fortin, J.A. (Eds.), Springer-Verlag, Heidelberg.

Dalpe, Y. (2001). *In vitro* monoxenic culture of arbuscular mycorrhizal fungi: a major tool for taxonomical studies. In: *Proceedings of the 3rd National Symposium on Mycorrhizal symbioses.* Guanajuato, Mexico, October, 2000. V. Olalde (Eds). Comite Nacional de Investigacion y enseñanza de la micorriza.

Dalpe, Y. and Declerck, S. (2002). Development of *Acaulospora rehmii* spore and hyphal swellings under root-organ culture. *Mycologia.* **94**: 850–855.

De Souza F.A. and Declerck, S. (2003). Mycelium development and architecture, and spore production of *Scutellospora reticulata* in monoxenic culture with Ri T-DNA transformed carrot roots. *Mycologia.* **95**: 1004-1012.

Declerck S., D'Or D., Cranenbrouck, S. and Le Boulenge, E. (2001). Modelling the sporulation dynamics of arbuscular mycorrhizal fungi in monoxenic culture. *Mycorrhiza.* **11**: 225–230.

Declerck, S., IJdo, M., Fernandez, K., Voets, L. and de la Providencia, I. (2009). Method and system for *in vitro* mass production of arbuscular mycorrhizal fungi. WO/ 2009/090220.

Declerck, S., Strullu, D.G. and Plenchette, C. (1996). *In vitro* mass-production of the arbuscular mycorrhizal fungus, *Glomus versiforme*, associated with Ri T-DNA transformed carrot roots. *Mycol. Res.* **100**: 1237–1242.

Declerck, S., Strullu, D.G., Plenchette, C. and Guillemette, T. (1996b). Entrapment of *in vitro* produced spores of *Glomus versiforme* in alginate beads: *in vitro* and *in vivo* inoculum potentials. *J. Biotechnol.* **48**: 51-57.

Declerck, S., Stullu, D.G. and Plenchette, C. (1998). Monoxenic culture of the intraradical forms of *Glomus* sp. isolated from a tropical ecosystem: a proposed methodology for germplasm collection. *Mycologia* **90**: 579–585.

Declerck, S. and Angelo-van Coppenolle, M.G. (2000). Cryopreservation of entrapped monoxenically produced spores of an arbuscular mycorrhizal fungus. *New Phytol.* **148**: 169–176.

Declerck, S., Cranenbrouck, S., Dalpe, Y., Granmougin-Ferjani, A., Fontaine, J. and Sancholle, M. (2000). *Glomus proliferum* sp. nov: a description based on morphological, biochemical, molecular and monoxenic cultivation data. *Mycologia.* **92**: 1178–1187.

Declerck, S., D'Or, D., Cranenbrouck, S. and Leboulengé, E. (2001). Modelling the sporulation dynamics of arbuscular mycorrhizal fungi in monoxenic culture. *Mycorrhiza.* **11**: 225–230.

Declerck, S., Strullu, D.G. and Plenchette, C. (1996). *In vitro* mass production of the arbuscular mycorrhizal fungus *Glomus versiforme* associated with Ri T-DNA transformed carrot roots. *Mycol. Res.* **100**: 1237–1242.

Declerck, S., Strullu, D.G. and Plenchette, C. (1998). Monoxenic culture of the intraradical forms of *Glomus* sp. isolated from a tropical ecosystem: a proposed methodology for germplasm collection. *Mycologia.* **90**: 579–585.

Diop, T. A. (1995). Ecophysiologie des champignons mycorhiziens à vésicules et arbuscules associés à Acacia albida Del. dans les zones sahélienne et soudano-guinéenne du Sénégal. Thèse de doctorat, Université d'Angers, France.

Diop, T. A., Plenchette, C. and Strullu, D. G. (1994). Monoxenic axenic culture of sheared-root inocula of vesicular-arbuscular mycorrhizal fungi associated with tomato roots. *Mycorrhiza* **5**: 17–22.

Diop, T.A. (2003). *In vitro* culture of arbuscular mycorrhizal fungi: advances and future prospects. *Afr. J. Biotechnol.* **2**: 692-697.

Douds, D. and Becard, G. (1993). Competitive interaction between *Gigaspora margarita* and *Gigaspora gigantia in vitro*. In: *Proceedings of the Ninth North American Symposium on Mycorrhizae*, Guelph, ON, Canada, August 8-12.

Douds, D.D., Jr. (1997). A procedure for the establishment of *Glomus mosseae* in dual culture with Ri T-DNA-transformed carrot roots. *Mycorrhiza.* **7**: 57–61.

Douds, D.D., Jr., and Schenck, N.C. (1990). Cryopreservation of spores of vesicular–arbuscular mycorrhizal fungi. *New Phytol.* **115**: 667–674.

Douds, D.D., Pfeffer, P.E. and Shachar-Hill, Y. (2000). In *Arbuscular mycorrhizas: physiology and function.* (Eds). Kapulnik, Y., D.D. Douds and J. Kluwer, Dordrecht. pp. 107–129.

Fontaine, J., Grandmougin-Ferjani, A., Glorian, V. and Durand, R. (2004). 24-Methyl: methylene sterols increase in monoxenic roots after colonization by arbuscular mycorrhizal fungi. *New Phytol.* **163**: 159–167.

Fortin, J.A., Becard, G., Declerck, S., Dalpe, Y., St-Arnaud, M., Coughan, A.P, Piche, Y. (2002). Arbuscular mycorrhiza on root-organ cultures. *Can. J. Bot.* **80**: 1-20.

Fortin, J.A., St-Arnaud, M., Hamel, C., Chaverie, C. and Jolicoeur, M. (1996). Aseptic *in vitro* endomycorrhizal spore mass production. US Pat.No. 5554530.

Gadkar, V., Adholeya, A. and Satyanarayana, T. 1997. Randomly amplified polymorphic DNA using the M13 core sequence of the vesicular-arbuscular mycorrhizal fungi *Gigaspora margarita* and *Gigaspora gigantea*. *Can. J. Microbiol.* **43**: 795–798.

Gianinazzi-Pearson, V., Branzanti, B. and Gianinazzi, S. (1989). *In vitro* enhancement of spore germination and early hyphal growth of vesicular-arbuscular mycorrhizal fungus by host root exudates and plant flavonoids. *Symbiosis.* **7**: 243-255.

Gilmore, A. E. (1968). Phycomycetous mycorrhizal organisms collected by open-pot culture methods. *Hilgardia.* **39**: 87–105.

Gryndler, M., Hrselova, H., Chvatalova, I. and Vosatka, M. (1998). *In vitro* proliferation of *Glomus fistulosum* intraradical hyphae from mycorrhizal root segments in maize. *Mycol. Res.* **102**: 1067–1073.

Hepper, C, M. and Jacobsen, I. (1983). Hyphal growth from spore of the mycorrhizal fungus *Glomus caledonius:* effects of amino-acids. *Soil Biol. Biochem.* **15**: 55-58.

Hepper, C.M. (1983). Limited independent growth of a vesicular arbuscular mycorrhizal fungus *in vitro. New Phytol.* **93**: 537-542.

Hepper, C. (1981). Techniques for studying the infection of plants by vesicular-arbuscular mycorrhizal fungi under axenic conditions. *New Phytol.* **88**: 641–647.

Hildebrandt, U., Janetta, K. and Bothe, H. (2002). Towards growth of arbuscular mycorrhizal fungi independent of a plant host. *Appl. Environ. Microbiol.* **68**: 1919-1924.

Jolicoeur, M. (1999). Optimisation d'un procédé de production de champignons endomycorhiziens en bioréacteur. Dissertation, École Polytechnique de Montréal.

Jolicoeur, M., Williams, R.D., Chavarie, C., Fortin, J.A. and Archambault, J. (1994). Production of *Glomus intraradices* propagules, an arbuscular mycorrhizal fungus, in an airlift bioreactor. *Biotechnol Bioeng.* **63**: 224–232.

Kandula, J., Stewart A. and Ridgway, H.J. (2006). Monoxenic culture of the arbuscular mycorrhizal fungus *Scutellospora calospora* and Ri T-DNA transformed carrot roots. *N. Z. Plant Prot.* **59**: 97-102.

Karandashov, V.E., Kuzourina, I.N., Hawkins, H.J. and George, E. (2000). Growth and sporulation of the arbuscular mycorrhizal fungus *Glomus caledonium* in dual culture with transformed carrot roots. *Mycorrhiza.* **10**: 23–28.

Koffi, M.C., de la Providencia, *I.E.*, Elsen, A. and Declerck, S. (2009). Development of an *in vitro* culture system adapted to banana mycorrhization. *Afr. J. Biotechnol.* **8**: 2750–2756.

MacDonald, R.M. (1981). Routine production of axenic vesicular arbuscular mycorrhizas. *New Phytol.* **89**: 87–93.

Mathur, N. and Vyas, A. (1995). *In vitro* production of *Glomus deserticola* in association with *Ziziphus nummularia. Plant Cell Rep.* **14**: 735–737.

Miller-Wideman, M.A. and Watrud, L. (1984). Sporulation of *Gigaspora margarita* in root culture of tomato. *Can. J. Microbiol.* **30**: 642–646.

Moses, B. (1988). Some studies relating to "independent" growth of vesicular-arbuscular endophytes. *Can. J. Bot.* **66**: 2533-2540.

Mosse, B. (1962). The establishment of vesicular–arbuscular mycorrhiza under aseptic conditions. *J. Gen. Microbiol.* **27**: 509–520.

Mosse, B. and Hepper, C.M. (1975). Vesicular–arbuscular infections in root-organ cultures. *Physiol. Plant Pathol.* **5**: 215–223.

Mosse, B. and Phillips, J.M. (1971). The influence of phosphate and other nutrients on the development of vesicular-arbuscular mycorrhiza in culture. *J. Gen. Microbiol.* **69**: 157–166.

Mugnier, J. and Moses, B. (1987). Vesicular-arbuscular infections in Ri T-DNA transformed roots grown axenically. *Phytopathololgy*. **77**: 1045-1050.

Nielsen, J.S., Joner, E.J., Declerck, S., Olsson, S. and Jakobsen, I. (2002). Phospho-imaging as a tool for visualization and noninvasive measurement of P transport dynamics in arbuscular mycorrhizas. *New Phytol*. **154**: 809-820.

Nogales, A., Camprubi, A., Estaun, V., Marfa, V. and Calvet, C. (2010). *In vitro* interaction studies between *Glomus intraradices* and *Armillaria mellea* in vines. *Spanish J. Agr. Res*. **8**: 62-68.

Pawlowska, T.E., Douds, D.D. and Charvat, I. (1999). *In vitro* propagation and life cycle of the arbuscular mycorrhizal fungus *Glomus etunicatum*. *Mycol. Res*. **103**: 1549–1556.

Pawlowska, T.E. and Taylor, J.W. (2004). Organization of genetic variation in individuals of arbuscular mycorrhizal fungi. *Nature*. **427**: 733–737.

Schreiner, R.P. and Koide, R.T. (1993). Mustards, mustard oils and mycorrhizas. *New Phytol*. **123**: 107–113.

Smith, S.A. and Read, D. (1997). Mycorrhizal symbiosis. (Ed 2nd).Academic Press, Inc., San Diego. Springer-Verlag, Heidelberg, pp. 341-375.

St-Arnaud, M., Hamel, C., Vimard, B., Caron, M. and Fortin, J.A. (1996). Enhanced hyphal growth and spore production of the arbuscular mycorrhizal fungus *Glomus intraradices* in an *in vitro* system in absence of host roots. *Mycol. Res*. **100**: 328–332.

Strullu, D.G., Diop, T.A. and Plenchette, C. (1997). Realisation de collections *invitro* de *Glomus intraradices* (Schenk et Smith) et de *Glomus versiforme* (Karsten et Berch) et proposition d'un cycle de développement. C. R. *Acad. Sci. Paris*, **320**: 41-47.

Strullu, D.G. and Romand, C. (1986). Méthode d'obtention d'endomycorhizes à vésicules et arbuscules en conditions axéniques C.R. *Acad. Sci. Ser. III Sci. Vie*, **303**: 245–250.

Strullu, D.G., Perrin, R., Plenchette, C. and Garbaye, J. (1991). Les mycorhizes des arbres et des plantes cultivées. Lavoisier, Paris.

Tiwari, P. and Adholeya, A. (2003). Host dependent differential spread of *Glomus intraradices* on various Ri T-DNA transformed root in vitro. *Mycol Prog*. **2**: 171–177.

Tiwari, P. and Adholeya, A. (2002). *In vitro* co-culture of two AMF isolates *Gigaspora margarita* and *Glomus intraradices* on Ri T-DNA transformed roots. *FEMS Microbiol. Letters* **206**: 39–43.

Voets, L., de la Providencia, *I.E.*, Fernandez, K., IJdo, M., Cranenbrouck, S. and Declerck, S. (2009). Extraradical mycelium network of arbuscular mycorrhizal fungi allows fast colonization of seedlings under *in vitro* conditions. *Mycorrhiza* **19**: 347–356.

Voets, L., Dupré de Boulois, H., Renard, L., Strullu, D.G. and Declerck, S. (2005). Development of an autotrophic culture system for the *in vitro* mycorrhization of potato plantlets. *FEMS Microbiol. Lett*. **248**: 111-118.

White, P.R. (1943). *A Handbook of Plant Tissue Culture*.

Chapter 3

Biocontrol of Soil-borne Diseases by Arbuscular Mycorrhizal Fungi

***Neeraj*[1]* *and Kanchan Singh*[1]**

ABSTRACT

Arbuscular mycorrhizal fungi are beneficial to the growth and health of soils and plants. Especially, for developing countries mycorrhizae provide an avenue to reduce the use of chemical fertilizers and its harmful effects on agricultural crops. PGPRs are another dominant group of rhizosphere bacteria which help in translocation, mobilization, solubilization of essential plant nutrients either in fixed or organic forms and thus, make them available to growing plants. On the basis of results of green house pot experiments, the experiments were carried out under field conditions to control the root-rot pathogen *Rhizoctonia solani* infecting *Phaseolus vulgaris* testing two arbuscular mycorrhizal fungi and a plant growth promoting rhizobacterium (PGPR)- *Pseudomonas fluorescens* (Pf) as biocontrol agents. These were inoculated in soil either alone or in combination before seed sowing and resulted in the improved plant-growth and reduced the root-rot disease severity. Both the AM fungi induced better plant growth and decreased disease incidence as compared to diseased control plants. *P. fluorescens* also proved beneficial for disease control. AMF+Pf combinations gave better

---

1 Botany Department, Feroze Gandhi College, Rae Bareli – 229001, India
*E-mail: aj_neer1@rediffmail.com

results against root-rot disease and also facilitated the AMF colonization in plant roots resulting in the increased plant growth and biomass.

Keywords: *Biocontrol, Arbuscular mycorrhiza, AM, AM fungi, PGPR, Pseudomonas, Phaseolus vulgaris.*

Introduction

Arbuscular mycorrhizal fungi (AMF) enter into beneficial intimate, symbiotic relationships with about 85 per cent of vascular plants (Parniske, 2008). Recent findings indicate that inheritable endobacteria inhabiting hyphae and spores of AM fungi (Naumann *et al.*, 2010) affect the metabolic profile of its host fungus. Synergizing with the rhizosphere bacteria- including plant growth-promoting rhizobacteria (PGPR) AMF act as efficient, dependable symbionts which can be used for better nutrient uptake, water relations, above-ground productivity for sustainable agriculture. These can also act as bio-protectants against pathogens and toxic stresses. Soil-plant-microbe interactions are complex and influence the plant health, productivity and soil fertility. Mycorrhizal fungi increase bio-control potential of *Pseudomonas fluorescens* (Siasou *et al.*, 2009).

Vesicular-arbuscular mycorrhizae have been reported to have a suppressive effect on diseases caused by a number of root infecting fungi (Dehne, 1982; Sharma *et al.*, 1992; 2008; Neeraj and Singh, 2010). Several investigators (Schenck and Kellam, 1978; Schonbeck, 1979) reported that the host plant previously inoculated with AM fungal symbiont exhibited increased resistance to several root-rots and wilts. Presence of AM fungi in soil was also reported to reduce plant root disease symptoms and pathogen population in soil (Linderman, 1994). The mycorrhizal roots are also known to enhance lignification and vasculature, which serve as a barrier (Dehne, 1982). The mycorrhizal symbioses promote excessive branching and this denser branched root system is less favourable for pathogen and nematodes, as they prefer primary roots (Stoffelen, 2000). AM fungi might also play a critical role in mitigating interactions between phytophagus insects (Gange, 2001). Increased nutrient uptake by mycorrhizal association results in more vigorous plants, acquisition of better tolerance or resistance to pathogen attack (Elsen *et al.*, 2001). This enhanced plant development may lead to disease escape or to higher tolerance against soil-borne pathogens (Dehne, 1982).

Chemical control of the pathogen is insufficient, but the presence of arbuscular mycorrhiza in plant roots is known to reduce the disease incidence (Kjoller and Rosendahl, 1996). It is well known that AM fungi can enhance plant growth (Bayozen and Yildiz, 2009), alter the cellular biochemical compositions (Neeraj and Singh, 2008; Jaiti *et al.*, 2008) and suppress plant diseases (Zachee *et al.*, 2008). These can increase the soil bacterial density, produce siderophores and reduce the nematode population in the mycorrhizosphere of plants. As compared to traditional application of chemical fertilizers these fungal symbionts enforce integrated plant nutrition management system with or without other treatments (Phirke *et al.*, 2008). Thirty-nine endophytic bacterial strains isolated from the nodule of *Lespedeza* sp. showed multiple plant growth promoting activity *i.e.*, indole acid production, ACC deaminase activity, siderophore production and phosphate solubilization (Palaniappan *et al.*, 2010). Establishment of *Glomus mosseae* in the root system effectively suppressed damping-off disease of chilli (Kavitha *et al.*, 2004). *Glomus mosseae* was also effective in reducing disease symptoms produced by *Phytophthora parasitica* in tomato plants (Pozo *et al.*, 2002). It is suggested that the higher specific phosphorus uptake in *Verticillium* inoculated pepper plants associated with *Glomus deserticola* could contribute to diminish the deleterious effect of pathogen on yield (Garmendia *et al.*, 2004). *G. fasiculatum* and *G. mosseae* have been reported to reduce the effect of *Fusarium oxysporum f. sp. lycopersici* on tomato and *Rhizoctonia solani* and *Pythium ultimum* on Poinsettia (Schenck and Kellam, 1978). Tahmatsidou *et al.* (2006) also reported that strawberry plants treated with AMF or PGPR showed significant reduction of wilt disease caused by *Verticillium dahliae*. Fusconi *et al.* (1999) found that AM fungus protected the root apices from the pathogenic infection, allowing normal root growth. Furthermore, larger apices which produce thicker roots, might indirectly contribute to plant protection. To make the things clear, we are presenting a case study first and then we will also discuss the control of other plant pathogens.

Control of Root-Rot by AM Fungi: A Case Study

The Host and the Pathogen

Phaseolus vulgaris L., commonly known as Rabi-Rajma, is an important crop cultivated in the Indo-Gangetic plains of India which suffers from root–rot disease caused by *Rhizoctonia solani* infection inflicting severe losses to plant growth and yield. Being soil-borne, below-ground infection, use of fungicidal application is invariably less effective for controlling the root-rot disease. Hence, the effect of applications of various combinations of AMF and *Pseudomonas fluorescens* as biological control agents along with an abiotic organic component *i.e.* mustard oil cake on *R. solani* infected *P. vulgaris* was studied. *P. vulgaris* L. *var* Amber 96-4 obtained from Indian Institute of Pulses Research (ICAR), Kanpur was used as host plant for different inoculants. *R. solani* strain (Rs; ITCC no. 2070) obtained from IARI, New Delhi was cultured on PD broth, homogenized and diluted to 10^6 mycelium bits/ml concentration.

The AM Fungi and PGPR

Soil based cultures of two AM fungi *Gigaspora albida* (Gia) and *Glomus sinuosum* (Gs) consisting of 250- 300 spores per 100 g soil and chopped, colonized root fragments were used for inoculations. The starter culture of a PGPR *Pseudomonas fluorescens* (Pf) was obtained from NBAIM, ICAR, Mau (India) was multiplied in KB-broth. Commercially available mustard oil cake (MOC) was air dried, powdered and added @10g/8 kg soil. MOC contains high amounts of glucosinolates, which is important for disease suppression.

The Experimental Design

A randomized block design field experiment was laid out in loam soil (pH 7.4), containing low organic carbon (0.27 per cent), nitrogen (0.44 per cent) and medium available phosphorus (24.2 kg/ha). Growth parameters *viz.* root, shoot lengths and fresh and dry weights of roots, shoots and leaves were recorded at regular intervals of 15 days from 30th to 105th days after sowing (DAS). Average pod number, pod length, pod dry weight per plant, average number of seed/plant and average 100-seed weight were also recorded.

Various combinations of AM Fungi studied were:

1. AMF* 6. AMF + Pf + Rs
2. AMF + Rs 7. AMF + Pf + Rs + MOC
3. Pf 8. Rs (Diseased Control)
4. Pf+ Rs 9. Untreated Control
5. AMF + Pf *Gs/Gia

Results and Discussion

The remarkable effects of different microbial treatments on the growth of diseased and disease-free Rajma plants as observed on 75th day after sowing are presented in Figures 3.1 and 3.2.

Dual inoculation of Gs and Pf along with MOC was best combination to control the root-rot disease of *P. vulgaris* caused by Rs and increasing the plant growth and yield. In the diseased plants this treatment produced 32.8 cm root length, 58.0 cm shoot length and 2.22g, 6.68g, 7.84g root, shoot and leaf dry weights respectively, followed by Gia + Pf + MOC treatment. Gs alone treatment increased lengths of roots (25.0 cm) and shoots (50.8 cm) as well as dry weight contents of roots (1.25 g), shoots (5.21 g) and leaves (5.52 g) as compared to control plants. Treatment with Gs alone proved better than *P. fluorescens* alone treatment and untreated control plants.

Effect of different treatments of *Glomus sinuosum* and *Gigaspora albida* on root, shoot and leaf dry weights of *P. vulgaris* 105 days after sowing are shown in Figures 3.3 and 3.4. The respective root, shoot and leaf dry weights per plant of diseased control were 0.90g, 2.59g and 3.47g while for un-inoculated control these biomasses were 1.02g, 3.84g and 4.45g. The GS + Pf + Rs + MOC treatment raised these biomasses respectively to 2.22g, 3.68g and 7.84g while for same inoculations without MOC these values were 1.90g, 6.36g and 7.27g. For Gia + Pf + Rs + MOC the respective dry weights for root, shoot and leaves per plant after 105 days were 2.19g, 6.62g and 6.74g respectively.

For any AMF in general, pod length and pod dry weight per plant were highest in AMF + Pf + MOC + Rs treated plants followed by AMF + Pf + Rs (Table 3.1). Pod dry weight per plant was minimum in Rs treatment; lesser than uninfected control plants. Comparing the two AM fungi, Gs with Pf + MOC + Rs was more effective for plant yield, which shows 6.8 cm pod length and 7.10 g pod dry weight. Thus, inoculation of an AMF in combination with *P. fluorescens* and mustard oil cake emerged as a good supporting bio-control system against the root-rot disease besides increasing the plant biomass and yield.

The root colonizations by AM fungi were also increased in plants inoculated with AMF + Pf + MOC + Rs as compared to plants inoculated with AMF + PF + Rs. The later treatment was than the plants treated with AMF + PF or AMF + Rs. In triple inoculated plants the root colonizations were 69 per cent with Gs and 65 per cent with Gia, respectively. The colonization in diseased dual inoculated (Gs+Rs) was 64 per cent whereas it was 61 per cent in (Gia+Rs) treatment which was higher than any AMF+Pf treated plant roots. The roots inoculated with AMF alone had only 55 per cent and 53 per cent colonization due to Gs and Gia respectively.

Table 3.1: Effect of Different Microbial Treatments on the Yield of *Phaseolus vulgaris* L.

Treatments	*Pod No. per Pt**	*Pod Length per Pt* (cm)*	*Pod Fresh Wt.* (g)*	*Pod Dry Wt.* (g)*	*Seed No. per Pt*	*100 Seed Weight (g)*
Gs	16.3	6.3	33.1	5.0	55.9	38.1
Gs + Rs	18.3	7.8	33.7	5.4	76.5	45.1
Gs + Pf	**20.3**	**9.2**	**36.6**	**6.9**	**95.1**	**51.3**
Gs + Pf + Rs	19.7	8.5	36.3	6.6	88.7	48.3
Gs + Pf + Rs + MOC	**21.3**	**9.5**	**36.9**	**7.1**	**100.8**	**54.7**
Gia	15.7	6.1	33.0	4.9	52.1	37.3
Gia + Rs	17.7	7.6	33.5	5.2	73.7	44.3
Gia + Pf	**20.3**	**8.8**	**36.3**	**6.7**	**93.1**	**50.7**
Gia + Pf + Rs	19.7	8.5	35.8	6.3	90.5	48.1
Gia + Pf + Rs + MOC	**20.7**	**9.4**	**36.8**	**7.1**	**100.2**	**53.1**
Pf	16.1	7.7	33.1	5.1	54.6	38.1
Pf + Rs	17.3	6.3	32.7	4.8	75.3	44.3
Rs	**13.7**	**4.5**	**26.3**	**2.1**	**39.5**	**30.7**
Control	**15.7**	**5.3**	**30.1**	**3.7**	**47.1**	**34.7**

Gs: *Glomus sinuosum*; Gia: *Gigaspora albida*; Rs: *Rhizoctonia solani*; Pf: *Pseudomonas fluorescens*; MOC: Mustard Oil Cake.

AMF+Rs and AMF+Pf treated plants had higher root colonization and plant growth than AMF alone inoculated plants. Similar to our results, Deene *et al.* (2004), observed that AMF association in plants had a detrimental influence on the incidence of wilt disease as compared to plants infected with *Fusarium udum* alone. They attributed this due to the enhancement of lignification and increase in the amount of vasculature in mycorrhizal roots, which served as barrier. Control of Rs infection can be attributed to *P. fluorescens* which reportedly increases antibiotic production in the presence of both AMF and pathogen (Siasou *et al.*, 2009). Further, organic amendments are also reported to increase the population of *P. fluorescens* which was found effective to reduce the bacterial wilt of tomato caused by *Ralstonia solanacearum* (Bora and Deka, 2007). *P. fluorescens* inoculations also resulted in improved seed germination, reduced wilting and improved biomass due to production of antifungal metabolites suppressing the root-rot disease. Besides accelerating the root growth, plant growth promoting rhizobacteria are also reported to considerably enhance mycorrhizal colonization (Neeraj and Singh, 2010). The beneficial antagonistic bacteria and fungi

Figure 3.1: Plant Biomass of *P. vulgaris* Increase Several Folds 75 Days After Sowing Due to Different Microbial Treatments along with the AM Fungus *Gigaspora albida* (Scale- 30 cm)

Figure 3.2: Plant Biomass of *P. vulgaris* Increase Several Folds 75 Days After Sowing Due to Different Microbial Treatments along with the AM Fungus *Glomus sinuosum* (Scale 30 cm)

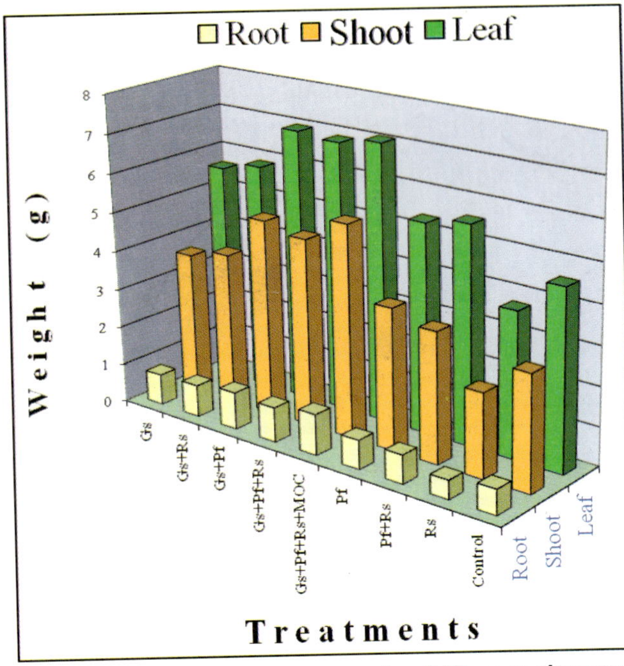

Figure 3.3: Effect of Different Treatments of *Glomus sinuosum* on Root, Shoot and Leaf Dry Weights of *P. vulgaris* 105 Days After Sowing

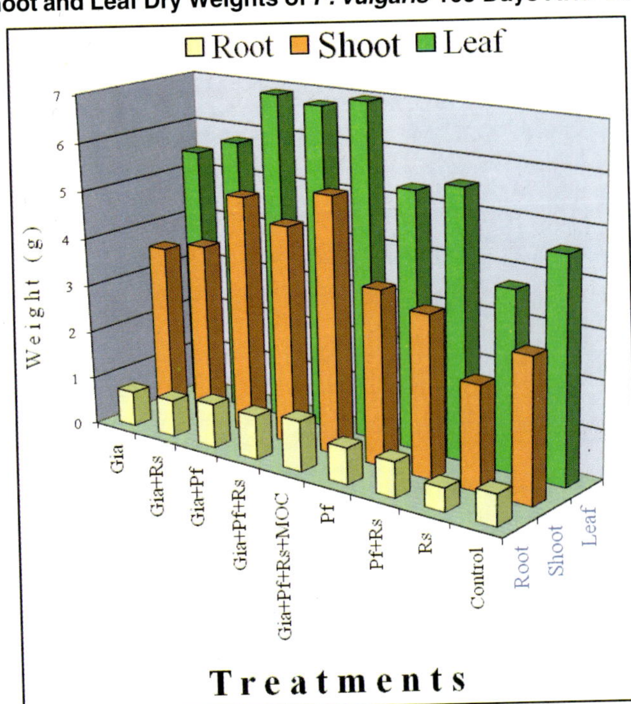

Figure 3.4: Effect of Different Treatments of *Gigaspora albida* on Root, Shoot and Leaf Dry Weights of *P. vulgaris* 105 Days After Sowing

compete with plant pathogens for nutrients and space by producing antibiotics, by parasitizing pathogens, or by inducing resistance in the host plants and therefore, these have been used for biocontrol of pathogens. Toussaint *et al.* (2008) also reported that *G. mosseae* conferred a bio-protective effect against *Fusarium oxysporium f. sp. basilica* as a result of increase in leaf rosmarinic and caffeic acid or essential soil concentrations. Systemic induction of many genes was predicted to be involved in stress or defense responses suggesting that mycorrhizal plants might display enhanced disease resistance (Liu *et al.*, 2007).

Control of other Plant Diseases through AMF

Many rhizosphere fungi, including mycorrhizal fungi, are able to suppress soil-borne plant pathogens (Whipps, 2001). Fungi have the advantage over bacterial biocontrol agents in that they are generally more effective at spreading through the soil and rhizosphere. A variety of mechanisms are involved in the control of fungal pathogens by rhizosphere fungi, including competition for nutrients, antibiotic production, and induced resistance. In addition, many fungi are able to parasitize spores, sclerotia, or hyphae of other fungi, resulting in biocontrol. Mycoparasitism is initiated by host sensing, which is generally followed by directed growth towards it, recognition, penetration, and degradation. Production of a number of degradative enzymes, including chitinases, proteases, and glucanases is involved in the biocontrol process. Although many agrochemicals are available to control root disease, concerns over the environmental and human health implications of pesticide use, and the desire for sustainable agricultural systems, have driven great interest in developing biological disease control based on rhizosphere antagonists of pathogens (Whipps, 2001). A number of such biocontrol agents are available as commercial formulations for use in agriculture and horticulture.

Inoculation with AM fungi and other soil microorganisms can affect both quantitative and qualitative microbial communities in the plant rhizosphere (Vazquez *et al.*, 2000). Bhawmik and Singh, (2004) showed that the plant growth promoting rhizobacteria considerably enhanced mycorrhizal colonization. They not only stimulated AM development, but also accelerated the root growth. The percentage of root colonization, plant height and dry matter production was increased in mycorrhizal plants treated with *Fusarium*. Thus, VAM association in plants had a detrimental influence on the incidence of wilt disease as compared to plants infected with *F. udum*. (Deene *et al.*, 2004). Root infection of barley by *Gaeumannomyces graminis var. tritici* was systemically reduced when barley plants showed high degrees of mycorrhizal root colonization, whereas a low mycorrhizal root colonization exhibited no effect on *G. graminis* infection (Khaosaad *et al.*, 2007). Mycorrhizae increase the P status of the host, which leads to a decrease in net leakage of root exudates and reduces pathogen activity (Graham and Menge, 1982). In leguminous plants *viz. Vigna mungo* and *Vigna radiata*, after dual inoculation with AM fungi and *Aspergillus niger*, root colonization increased in the treatment supplemented with rock phosphate. Plants inoculated with *Glomus fasciculatum* and *A. niger* showed greater concentration of phosphorus and nitrogen, leaf chlorophyll and total soluble sugars than the non-mycorrhizal plants (Rao and Rao, 1998). Rani *et al.* (1998) reported that dual inoculation with VAM and *Rhizobium* sp. resulted in maximum plant growth and nodulation. It was concluded that the VAM fungi in combination with *Rhizobium* was the best biofertilizer for biomass production of multipurpose *Acacia nilotica* plants. In sorghum plants *Streptomyces coelicolar* strain increased the intensity of mycorrhizal root colonization and arbuscule formation (Abdel-Fattah and Mohamedin, 2000). Healthy and infected plants of two varieties of sesame- Rajeshwari and Gowri exhibited different mycorrhizal colonization. In both the cultivars colonization was more in phyllody infected plants than the healthy plants. Vesicles and mycelium were associated with all types of plants. However, arbuscules were not found in healthy plants, through present in infected plants (Shailaja *et al.*, 1998). Ozgonem *et al.* (2010) reported that AMF could effectively be used against stem rot caused by *Scelrotium rolfsii*. The severity of disease was reduced 37.8–64.7 per cent due to AMF. Toussaint *et al.* (2008) also reported that *Glomus mosseae* could confer a bio-protective effect against *Fusarium oxysporium f. sp. basilica* as a result of increase in leaf rosmarinic and caffeic acid or essential oil concentrations.

Systemic induction of many genes was predicted to be involved in stress or defense response of the shoot of mycorrhizal *Medicago truncaluta* plants suggesting that mycorrhizal plants might display enhanced disease resistance (Liu *et al.*, 2007). Dual inoculations (*i.e.* use of two biological control entities other than the pathogen) have been found more effective in disease suppression than single inoculations (Kavitha *et al.*, 2003). Combined inoculation of AM fungi, phosphate solubilizing bacteria (PSB) *Bacillus polymixa* and *Azospirillum brasilense* resulted in maximum growth response both under fertilized and unfertilized conditions in two types of soil (Muthukumar and Udaiyan, 2006). Certain rhizosphere fluorescent *Pseudomonas* strains are engaged as potential biocontrol agents because of their ability to produce antibiotics. *Pseudomonas fluorescence* reportedly produced antibiotics only in presence of AMF or soil-borne pathogen and also this production increased in the presence of both AMF and pathogen (Siasou *et al.*, 2009). Shalaby and Hanna (1998) reported that *Glomus mosseae* combined with *Bradyrhizobium japonicum* increased the shoot dry weight, nodulation, accumulated nitrogen and phosphorus as well as nitrate reductase activity in soybean plants. Further, organic amendments increased population of *P. fluorescens* and decreased bacterial wilt caused by *Ralstonia solanacearum* of tomato (Bora and Deka, 2007). The maximum values of growth parameters were recorded in dual inoculations of biocontrol agents AMF+ *Pseudomonas fluorescens* with mustard oil cake (an organic material) against root-rot disease of French bean caused by *Rhizoctonia solani* (Neeraj and Singh, 2009). AM fungi are also found to be effective in the suppression of damping-off disease of chilli (Kavitha *et al.*, 2004) and *Phytophthora parasitica* infection of tomato plants (Pozo *et al.*, 2002). Dual inoculations have been found more effective in disease suppression than single inoculations (Kavitha *et al.*, 2003). Similarly inoculation of an arbuscular mycorrhizal fungi (AMF) in combination with *P. fluorescens* and mustard oil cake showed best supporting biocontrol system against the root- rot disease of *Cyamopsis tetragonaloba* besides increasing the plant height, weight and yield (Neeraj and Singh, 2010).

AMF plants had a higher net photosynthetic rate, higher leaf elemental N, P, and K, and generally greater growth than non-AMF plants. Total colonization levels of

AMF plants ranged from 27 per cent (100 per cent OSRF (organic slow release fertilizer) to 79 per cent (30 per cent OSRF). Root acid phosphatase (ACP) and alkaline phosphatase (ALP) activities were generally higher in AMF than non-AMF plants. AMF are also reported to improve the nutrient acquisition from an organic fertilizer source by enhancing root acid-phosphatase and alkaline-phosphatase activity and thus facilitating inorganic P acquisition, increasing photosynthesis, and improving plant growth (Amaya-Carpio *et al.*, 2009).

Establishment of *Glomus fasciculatum* and *Glomus mosseae* in the root system of host resulted in decrease in the final nematode population and significant plant growth enhancement (Sharma and Trivedi, 2000). Colonization of soybean roots by mycorrhizal fungi was negatively correlated with *Heterodera glycines* population densities due to nematode antagonism to the mycorrhizal fungi rather than suppression of nematode populations (Winkler *et al.*, 1994). *Glomus mosseae* with margosa cake significantly reduced root galling, egg mass production and produced vigorous tomato seedlings with increased root colonization by *G. mosseae* (Reddy *et al.*, 1998). Sankarnarayana and Sundarababu (1999) also reported that combination of super phosphate, VAM and the nematode recorded higher plant growth, minimum nematode population, higher VAM spore population and colonization. Therefore, co-inoculation of mycorrhizal fungi with PGPR and amendment with organic supplement-mustard oil cake, which has sulphur containing compound glucosinolates, isothiocyanates and nitriles that have been demonstrated to control fungi, bacteria and nematodes (Mojtahedi *et al.*, 1991; Neeraj and Singh, 2010) may provide and eco-friendly and sustainable strategy to control the root- rot losses.

The mycorrhizal fungi are able to provide protection to the host plant against root and shoot pathogens (Whipps, 2004). They might do this in a number of ways, including antibiotic production, induced resistance, competition for root infection sites, and by providing a physical barrier to infection. The significance and function of plant–mycorrhizal associations, while not as diverse as the plant–rhizosphere association, can vary greatly (Smith and Read, 1997). A single plant root may be colonized by many different mycorrhizal fungi, and mycorrhizal fungi often have low specificity, and are able to colonize a variety of different plant species. There is also great variation between fungal species in the benefits they provide to their host (Morgan *et al.*, 2005).

Many practices, such as crop rotation, seed certification, resistant cultivars, chemical fungicides, soil fumigation etc, are used to control plant pathogens, especially soil-borne pathogens. However, there are many problems associated with controlling pathogens with long-term persistent survival structures due to difficulties in reducing pathogen inoculum and lack of good sources of plant resistance. Soil fumigants *e.g.* methyl bromide is highly toxic and also depletes the stratospheric ozone layer (Gan *et al.*, 1997). Therefore, many researchers are trying to develop alternate approaches based on either adding or manipulating microorganisms to enhance plant protection against pathogens (Grosch *et al.*, 2005). Both, the seeds and soil can be treated. The beneficial microorganisms (antagonistic bacteria *e.g. Pseudomonas fluorescens*, *Bacillus subtilis* etc.) and fungi (*e.g.* AM fungi, *Trichoderma* spp. etc.) compete with plant pathogens for nutrients and space by producing antibiotics, by parasitizing pathogens, or by inducing resistance in the host plants. These have been found beneficial for biocontrol of pathogens (Berg *et al.*, 2007).

There are clearly ample evidences to show that AM fungi can act as biocontrol agents in a range of systems. Often this may go unnoticed in the environment when improved plant growth is the only measure of efficacy available but direct competition or inhibition, biochemical changes associated with plant defense mechanisms and induced resistance and development of an antagonistic micro biota may all be taking place. Soon with the application of molecular analytical techniques, the features controlling the biocontrol activity in mycorrhiza would be clearly understood. Thus, the association of AMF with PGPR will attract increasing attention as an unexploited resource in biological control. Their performance in the real world should be given increased attention through field testing and may provide an environmentally safe alternative to chemical pesticides.

References

Abdel-Fattah, G.M. and Mohamedin, A.H. (2000). Interaction between a vesicular-arbuscular mycorrhizal fungus (*Glomus intraradices*) and *Streptomyces coelicolor* and their effects on *Sorghum* plants grown in soil amended with chitin of brown scales. *Biology and Fertility of Soil*, **32**: 401-409.

Amaya-Carpio, L., Davis, F.T. and Fox, T. and He, C. (2009). Arbuscular mycorrhizal fungi and organic fertilizer influence photosynthesis, root phosphatase activity, nutrition and growth of *Ipomoea carnea* spp. fistula. *Photosynthetica*. **47**: 1-10.

Bayozen, A, and Yildiz, A. (2009). Determination of mycorrhizae interactions and pathogenicity of *Rhizoctonia solani* Kuhn isolated from strawberry and *Xanthium strumarium*. *Turkish J. Biol.* **33**: 53-57.

Berg, G., Grosch, R. and Scherwinski, K. (2007). Risk assessment for microbial antagonists: Are there effects on non-target organisms? *Gesunde Pflanzen* **59**: 107-117.

Bhowmik, S.N. and Singh, C.S. (2004). Mass multiplication of AM inoculum: Effect of plant growth promoting rhizobacteria and yeast in rapid culturing of *Glomus mosseae*. *Curr. Sci.* **86**: 705-709.

Bora, L. C. and Deka, S. N. (2007). Wilt disease suppression and yield enhancement in tomato (*Lycopersicon esculentum*) by application of *Pseudomonas fluorescens* based biopesticide (Biofor Pf) in Assam. *Indian J. Agric. Sci.* **77**: 490-494.

Deene, S., Garampalli, R. and Reddy, C. N. (2004). Influence of dual inoculation with vesicular- arbuscular mycorrhiza (*Glomus aggregatum*) and *Fusarium udum* on plant vigour and wilt disease incidence of pigeonpea. *Phytomorph.* **54**: 23-29.

Dehne, H.W. (1982). Interaction between vesicular-arbuscular mycorrhizal fungi and plant pathogens. *Phytopathology*. **72**: 1115-1119.

Elsen, A., Declerck, S. and De Waele, D. (2001). Effect of *Glomus intraradices* on the reproduction of the burrowing nematode (*Radopholus similis*) in dixenic culture. *Mycorrhiza*. **11**: 49-51.

Fusconi, A., Gnari, E., Trotta, A. and Berta, G. (1999). Apical meristems of tomato roots and their modifications induced by arbuscular mycorrhizal and soil borne pathogenic fungi. *New Phytol.* **142**: 505-516.

Gan, J., Yates, S.R., Spencer, W.F., Yates, M.V. and Jury, W.A. (1997). Laboratory scale measurements and simulations of effect of application methods on soil methyl bromide emission. *J. Environ. Quality.* **26**: 310-317.

Gange, A.C. (2001). Species-specific responses of a root and shoot feeding insect to arbuscular-mycorrhizal colonization of its host plant. *New Phytol.* **150**: 611-618.

Garmendia, I., Goicoechea, N. and Aguirreolea, J. (2004). Effectiveness of three *Glomus* species in protecting pepper (*Capsicum annum* L.) against *Verticillium* wilt. *Biological Control.* **31**: 296-305.

Graham, J.H. and Menge, J.A. (1982). Influence of vesicular- arbuscular mycorrhiza and soil phosphorus on take- all disease of wheat. *Phytopathology.* **72**: 95-98.

Grosch, R., Lottmann, J., Faltin, F. and Berg, G. (2005). Use of bacterial antagonists to control diseases caused by *Rhizoctonia solani. Gesunde Pflanzen.* **57**: 199-205.

Jaiti, F., Kassami, M., Meddich, A., El Hadrami, I. (2008). Effects of arbuscular mycorrhization on accumulation of hydroxycinnamic acid derivatives in date palm seedlings challenged with *Fusarium oxysporium* f sp *albedinis. J. Phytopathol.* **156**: 641-646.

Kavitha, K., Kumari, M. and Siva Prasad, P. (2003). Effect of dual inoculation of native arbuscular- mycorrhizal fungi and *Azospirillum* on suppression of damping-off in chilli. *Indian Phytopath.* **56**: 112-113.

Kavitha, K., Kumari, M. and Siva Prasad, P. (2004). Arbuscular- mycorrhizal fungi for bio-control of damping-off in chilli. *J. Mycol. Plant Pathol.* **34**: 349-350.

Khaosaad, T., Garcia-Garrido, J.M., Steinlcellner, S. and Vierheilig, H. (2007). Take-all disease is systemically reduced in roots of mycorrhizal barley plants. *Soil Biol. Biochem.* **39**: 727-734.

Kjøller, R., Rosendahl, S. (1996). The presence of the arbuscular mycorrhizal fungus *Glomus intraradices* influences enzymatic activities of the root pathogen *Aphanomyces euteiches* in pea roots. *Mycorrhiza* **6**: 487-491.

Linderman, R.G. (1994). Role of VAM fungi in biocontrol, In: *Mycorrhizae and Plant Health,* APS Press, The American Phytopathological Society, St. Poul, Minnesota, p: 1-26.

Liu, J. Y., Maldonado-Mendoza, I., Lopez-Meyer, M., Cheung, F., Town, C. D. and Harrison, M. J. (2007). Arbuscular mycorrhizal symbiosis is accompanied by local and systemic alterations in gene expression and an increase in disease resistance in the shoots. *Plant J.* **50**: 529-544.

Mojtahedi, H., Santo, G.S., Hang, A.N. and Wilson, J.H. (1991). Suppression of root-knot nematode populations with selected rape seed cultivars as green manures. *J. Nematol.* **23**: 170-174.

Morgan, J.A.W., Bending, G. D. and White, P. J. (2005). Biological costs and benefits to plant–microbe interactions in the rhizosphere. *J Exptl. Bot.*, **56**: 1729–1739.

Muthukumar, T. and Udaiyan, K. (2006). Growth of nursery-grown bamboo inoculated with arbuscular mycorrhizal fungi and plant growth promoting rhizobacteria in two tropical soil types with and without fertilizer application. *New Forests*. **31**: 469-485.

Neeraj and Singh, K. (2008). Biochemical changes in *Phaseolus vulgaris* L. dual inoculated with arbuscular mycorrhizal fungi and *Rhizobium*. *Indian J. Bot. Res.* **4**: 73-88.

Neeraj and Singh, K. (2009). Impact of AM fungi along with *Pseudomonas fluorescens* and mustard oil cake in controlling root-rot of French bean. *Mycorrhiza News*. **21**: 12-14.

Neeraj and Singh, K. (2010). *Cyamopsis tetragonaloba* L. Taub inoculated with arbuscular mycorrhizae and *Pseudomonas fluorescens* and treated with mustard oil cake overcome *Macrophomina* root- rot losses. *Biol. Fertil. Soil*. **46**: 237- 245.

Naumann, M., Schussler, A., Bonfonte, P. (2010). The obligate endobacteria of arbuscular mycorrhizal fungi are ancient heritable components related to Mollicutes. *ISME Journal* **4**: 862-871.

Ozgonem, H., Akgul, D.S. and Erkilic, A. (2010). The effects of arbuscular mycorrhizal fungi on yield and stem rot caused by *Sclerotium rolfsii* sac. in peanut. *African J. Agri. Res.* **5**: 128-132.

Palaniappan, P., Chauhan, P.S., Saravanan, V.S., Anandham, R. and Sa, T. (2010). Isolation and characterization of plant growth promoting endophytic bacterial isolates from root nodule of *Lespedeza* sp. *Biol. Fertil. Soils* **46**: 807-816.

Parniske, M. (2008). Arbuscular mycorrhiza: the mother of plant root endosymbioses. *Nat. Rev. Microbiol.* **6**: 763-775.

Phirke, N.V., Kothari, R.M., Chincholkar, S.B. (2008). Rhizobacteria in rhizosphere improved plant health and yield of banana by offering proper nourishment and protection against plant diseases. *Appl. Biochem. Biotechnol.* **151**: 441-451.

Pozo, M.J., Cordier, C., Dumas-Gaudot, E., Gianinazzi, S., Barea, J.M. and Azcon-Aguilar, C. (2002). Localized versus systemic effect of arbuscular-mycorrhizal fungi on defence response to *Phytophthora* infection in tomato plants. *J. Exptl Bot.* **53**: 525-537.

Rani, P., Agarwal, A. and Mehrotra, R.S. (1998). Growth responses in *Acacia nilotica* inoculated with VAM fungus (*Glomus fasciculatum*), *Rhizobium sp.* and *Trichoderma harzianum*. *J. Mycopathological Res.* **36**: 13-16.

Rao, V.U. and Rao, A.S. (1998). Interactive effect of VAM fungi and *Aspergillus niger* on nutrient and biochemical constituents in two grain legumes. *Indian J. Tropical Agric.* **14**: 115-121.

Reddy, P.P., Nagesh, M., Divappa, V. and Kumar, M.V.V. (1998). Management of *Meloidogyne incognita* on tomato by integrating endomycorrhiza, *Glomus mosseae* with oil cakes under nursery and field conditions. *Zeitschrift fur Pflanzenkrankheiten and Pflanzenschutz* **105**: 53-57.

Shailaja, K.M., Pavankumar, P. and Reddy, S.R. (1998). Influence of phyllody disease on VAM colonization in two cultivars of Sesame. *Indian Phytopathol.* **51**: 273-274.

Sharma, A. and Trivedi, P.C. (2001). Concomitant effect of VAM and neem products on *Heterodera avenae* infected wheat. *J. Indian Bot Soc.* **80**: 9-13.

Schenck, N.C. and Kellam, M.K. (1978). The influence of vesicular arbuscular mycorrhiza and disease development. *Florida. Agric. Stn. Tech. Bull.* **798**: 16.

Schussler, A., Schwarzott, D. and Walker, C. (2001). A new fungal phylum, the Glomeromycota: phylogeny and evolution. *Mycol. Res.* **105**: 1413-1421.

Schonbeck, F. (1979). Endomycorrhizas in relation to plant disease. In: *Soil Borne Plant Pathogens*. Scheppers, B. and Gams, W. (eds.) pp: 271-280.

Shalaby, A.M. and Hanna, M.M. (1998). Preliminary studies on interaction between VA Mycorrhizal fungus *Glomus mosseae, Bradyrhizobium japonicum* and *Pseudomonas syringae* in soyabean plants. *Acta Microbiol. Polonica.* **47**: 385-391.

Sankaranarayanan, C. and Sundarababu, R. (1991). Effect of phosphatic fertilizers on the interaction of *Glomus mosseae* and *Meloidogyne incognita* on black gram (*Vigna mungo*). *Indian Nematol.* **29**: 44-47.

Sharma, A.K., Johri, B.N. and Gianianzzi, S. (1992). Vesicular arbuscular mycorrhizae in relation to plant disease. *World J. Microbiol. Biotechnol.* **8**: 559-563.

Sharma. S., Aggarwal, A. and Kumar, A. (2008). Bio-control of wilt of *Albizzia lebbek* by using AM fungi and *Trichoderma viride*. *Ann. Plant Protec. Sci.* **16**: 422-424.

Siasou, E., Standing, D., Killham, K., Johmson, D. (2009). Mycorrhizal fungi increase bio- control potential of *Pseudomonas fluorescens*. *Soil Biol. Biochem.* **41**: 1341-1343.

Smith, S.E. and Read, D.J. (1997). *Mycorrhizal Symbiosis*, 2nd edn. Academic Press, London, UK.

Stoffelen, R. (2000). Early screening of *Eumusa* and *Australimusa* bananas against root-lesion and root-knot nematodes. Dissertationes de Agricultura, Katholieka Universiteit, Leuven, Belgium, No. 426.

Tahmatsidou, V., Sullivan, J.O., Cassells,. A.C., Voyiatzis, D. and Paroussi, G. (2006). Comparison of AMF and PGPR inoculants for the suppression of *Verticillium* wilt of strawberry (*Fragaria ×ananassa C.V. Selva*). *Appl. Soil Ecol.* **32**: 316-324.

Toussaint, J.P., Kraml, M., Nell, M., Smith, S.E., Smith, F.A., Steinkellner, S., Schmiderer, C., Vierheilig, H. and Novak, J. (2008). Effect of *Glomus mosseae* on concentrations of rosmarinic and caffeic acids and essential oil compounds in basil inoculated with *Fusarium oxysporum f.sp. basilica*. *Plant Pathol.* **57**: 1109-1116.

Vazquez, M.M., Bejarano, C., Azcon, R. and Barea, J.M. (2000). The effect of a genetically modified *Rhizobium meliloti* inoculant on fungal alkaline phosphatase and succinate dehydrogenase activities in mycorrhizal alfalfa plants as affected by the water status in soil. *Symbiosis.* **29**: 49-58.

Whipps, J.M. (2004). Prospects and limitations for mycorrhizas in biocontrol of root pathogens. *Canadian J. Bot.* **82**: 1198-1227.

Winkler, H.E., Hetric, B.A.D. and Todd, T.C. (1994). Interactions of *Heterodera glycines, Macrophomina phaseolina* and Mycorrhizal fungi on soybean in Kansas. *Supplement to J. Nematol.* **26**: 675-682.

Zachee, A., Bekolo, N., Bime Dooh, N., Yalen, M., Godswill, N. (2008). Effect of mycorrhizal inoculum and urea fertilizer on disease development and yield of groundnut crops (*Arachis hypogaea* L.). *African J. Biotech.* **7**: 2823-2827.

Chapter 4

Comparative Analysis of Drinking Water Quality of Paharia and Non-Paharia Villages of Dumka District, Jharkhand, India

Rajendra Pandey[1]

ABSTRACT

The comparative analysis of drinking water of Dumka District of Santhal Pargana Division of Jharkhand State was carried out from January 2010 to December 2010. In the present study potable water for drinking purposes were collected from four different sources namely wells, tubewells, ponds and Mayurakshi river. Many important physico-chemical parameters were taken into consideration. During observation it has been detected that certain standard permissible limits as suggested by Indian Standard Institute (ISI) and World Health Organization (WHO) had crossed. The villagers suffered from a number of water borne diseases such as diarrohoea, gastritis, blood dysentery etc. owing to changes in the quality of water made for drinking purposes. It has concluded that paharias were worse sufferer in comparison to non-paharias.

Keywords: *Chemical, Paharia, Physicochemical, Quality of water.*

---

1 Department of Botany, S.K.M. University, Dumka – 814 101, Jharkhand, India

Introduction

Santhal Pargana is a hilly terrain of Jharkhand whose maximum population resides in the lap of nature. It lies between 86°28′ and 89° 57′ east longitude and 23° 48′ and 25° 18′ north latitude in Jharkhand. Dumka being the Head quarter of this division, lies at 25° 18′ north latitude and 87° 37′ east longitude. The villages which were chosen for investigation are situated 4 km. away from the Dumka Bus stand. Villagers of Santhal Pargana get their potable water from river (Mayurakshi), Ponds, Wells and Govt. tubewells. Two villages namely one paharia (Asansol) and the other non- paharia (Chorkatta) were selected for the present study. The very sources of potable water contains both micro and macro nutrients in permissible limit but quality of drinking water changes due to human interference and get contaminated through percolation and seepage, drains and domestic sewage. Thus the quality of potable water deteriorates leading to various health hazards such as diarrohoea, gastritis and blood dysentery. The present study is an attempt to assess the potable water quality of different water resources of the two villages.

Materials and Methods

Potable water samples were collected from different sources *viz.* Mayurakshi river, ponds, wells and Govt. tube-wells from the rural areas (villages). Temperature of water was recorded by mercury centigrade thermometer. The pH was measured by both pH paper in the field and by digital pH meter in the laboratory, free CO_2 and HCO_3- alkalinity were determined by the method suggested by Welch (1948). Dissolved oxygen was measured by Winkler's modified volumetric methods. Chloride and nitrate nitrogen were calculated by standard method of NEERI (1979). For the estimation of other chemicals constitutes of potable water standard methods of APHA (1980) and WHO (1985) were followed.

Results and Discussion

The physico- chemical parameters of paharia village (Asansol) and non- paharia village (Chorkatta) have been summarized in Tables 4.1 and 4.2 respectively. From Tables 4.1 and 4.2 it is evident that turbidity values of pond were maximum 23.5 NTU and 23.2 NTU in both the cases, whereas minimum was recorded in well 15.3 NTU in non- paharia village and 14.2 NTU in paharia village well respectively. The upper limit of ISI and WHO are 10 NTU and 5 NTU respectively. Total solids values were maximum in ponds and rivers in both the cases whereas minimum was found in Govt. tubewell. The higher value of turbidity and total solids of potable water may create taste, odour and colour problems. It may also cause throat and gastro-intestinal problems (ISI, 1983 and WHO, 1984a). The pH values were always found to be more than 7.0 except in the river (6.8). (Kumar and Saha, 1989)

The minimum value of dissolved oxygen was recorded in Govt. tubewell while maximum was in river and pond. The concentration of bicarbonate was maximum in Govt. tubewell while minimum in the river (128.0 ppm). The concentration of free CO_2 and HCO_3^-alkalinity were directly related to each other and their concentration varies according to the depth. (Saha and Pandey, 1987)

Table 4.1: Average Physico-Chemical and Bacteriological Quality of Different Water Sources of Paharia Village (Asansol)

Parameters	*Mayurakshi River*	*Pond*	*Well*	*Govt. Tube-well*	*Quality Criteria of Potable Water Prescribed by*	
					ISI	*WHO*
Water temps (C)	23.9	26.1	25.9	26.2		
Turbidity (NTU)	22.2	23.2	13.2	14.2	10.0	5.0
Total solids (ppm)	670.0	674.0	647.0	392.0	500	1500
pH	6.8	7.4	7.3	7.1	6.5	6.5
Dissolved O_2 (ppm)	5.6	5.9	1.8	1.9		
Free CO_2 (ppm)	2.7	2.9	17.7	12.7		
CO_3- alk (ppm)	38.2	40.2	Nil	Nil		
HCO_3- alk (ppm)	129.0	153.0	247.0	307.0		
Total Hardness (ppm)	94.0	95.5	103.5	325.5	300	500
Calcium Hardness (ppm)	58.5	60.1	66.8	102.5	75	200
Mg. Hardness (ppm)	38.2	37.7	36.5	137.7	30	50
Conductivity(mho)	260.0	262.0	678.0	715.0		
Phosphate(ppm)	0.06	0.07	0.06	0.08		
Chloride (ppm)	65.75	67.58	376.98	290.58	250	600
Nitrate (ppm)	2.127	4.129	20.16	16.75	45	10
Silicate (ppm)	35.55	37.18	19.27	71.17		
Sulphide (ppm)	BDL*	BDL*	BDL*			
Na^+ (ppm)	25.6	27.8	47.76	95.8		
K^+ (ppm)	0.16	1.18	2.60	3.75		
Total Bacterial density/L	0.144×10^7	1.750×10^7	1.565×10^7	0.490×10^7		
MPN Coliform/100ml	887	898	560	148		

* BDL: Below detection limit.

Govt. tube-well and wells showed maximum values of total hardness, chloride, nitrate, calcium, magnesium hardness, conductivity, phosphate, silicate, sodium and potassium while pond and river showed their minimum values. These values crossed the permissible limit (ISI, 1983; WHO, 1984a).

Chemically chloride is one of the most important pollution indicator of the water. Its higher concentration in well and Govt. tube-wells water show their sources to be more polluted than of others. Nitrate nitrogen concentration was maximum in wells followed by Govt. tube-well water (16.75). These values are more than the permissible limits of WHO, 1984a and hence may causes methaemoglobinaemia among the infants. (USEPA, 1977; WHO, 1984b)

Table 4.2: Average Physico-Chemical and Bacteriological Quality of Different Potable Water Sources of Non-Paharia Village (Chorkatta)

Parameters	*Mayurakshi River*	*Pond*	*Well*	*Govt. Tubewell*
Water temps (ºC)	23.9	24.7	27.0	29.0
Turbidity (NTU)	22.2	23.2	15.3	15.7
Total solids (ppm)	668.0	669.0	656.0	398.5
pH	6.8	7.3	7.2	7.0
Dissolved O_2 (ppm)	5.6	5.8	3.0	2.1
Free CO_2 (ppm)	2.7	2.6	18.6	14.0
CO_3- alk (ppm)	38.2	40.2	Nil	Nil
HCO_3- alk (ppm)	128.0	150.0	243.0	307.7
Total Hardness (ppm)	94.0	94.5	105.5	327.5
Calcium Hardness (ppm)	58.5	60.0	67.5	200.5
Mg. Hardness (ppm)	38.2	37.0	36.0	137.0
Conductivity(mho)	260.0	261.5	657.3	704.0
Phosphate(ppm)	0.06	0.07	0.07	0.08
Chloride (ppm)	65.75	66.50	428.58	318.50
Nitrate (ppm)	2.127	3.250	18.155	16.37
Silicate (ppm)	35.55	36.66	20.50	72.75
Sulphide (ppm)	BDL*	BDL*	BDL*	BDL*
Na^+ (ppm)	25.60	26.50	45.60	96.70
K^+ (ppm)	0.16	1.15	2.65	3.85
Total Bacterial density/L	0.144×10^7	1.546×10^7	1.457×10^7	0.397×10^7
MPN Coliform/100ml	887	895	558	140

* BDL: Below detection limit.

Potable water must be free from pathogenic and no-pathogenic microbes. Maximum bacterial density was recorded in well and pond water while minimum was in river and tube-well. The maximum value of MPN of coliform was recorded in river and pond while minimum was found in Govt. tube- well. Due to such contaminated potable water different types of water born diseases are quite common in the area. Presence of *Escherichia coli, Aerobacter aerogens, Salmonella typhii, Staphylococcus aureus* and *Clostridum perfriges* (*C. welchii*) shows the real pollution of potable water with faecal matter. *E. coli* in drinking water is responsible for causing enteritis in young children, and along with *A. aerogens*, it may cause diarrhoea in adults (Saha *et al.*, 1987).

After going through the data presented in Tables 4.1 and 4.2 it can be concluded that paharias are prone to several diseases in comparison to non-paharias due to consumption of pond and well water which they use as drinking water. If the adequate use of alum, bleaching powder and other disinfectants is made and villagers are made fully aware about hygienic use of potable water, the water borne diseases may be controlled. The Govt. must also take care of ecologically uneducated paharias to improve their potable water quality for drinking and other purposes.

Acknowledgements

We are grateful to Dr. N.K. Thakur and Dr. S.N. Jha and others for providing facilities and help in identifying the microbes.

References

APHA (1980). *Standard Methods for the Examination of Water and Wastewater*, 15th edn. American Public Health Association, New York, p. 1134.

ISI (1983). *Indian Standards: Specifications for Drinking Water*. Indian Standard Institute, New Delhi, p. 22.

Kumar, Sheo and Saha, L.C. (1989). Assessment of drinking water quality of Bhagalpur. *Biol. Bul. of India*. **11**: 9-13.

NEERI (1979). *A Course Manual: Water and Wastewater Analysis*. National Environmental Engineering Research Institute, Nagpur, p. 134.

Saha, L.C and Pandey, B.K. (1987). Quality of hand pump waters at Bhagalpur, I. Bacteriological quality. *Acta. Ecol.*, **9**: 44-48.

Saha. L.C., Pandit, B. and Pandey, B.K. (1987). Bhagalpur well water. Bacteriological quality. *Nat. Acad. Sci. Letters*, **10**: 311-313

USEPA (1977). National interim primary drinking water regulation. *United States Environmental Protection Agency, U.S.A.*, p. 159.

Welch, P.S. (1948). *Limnological Methods*. Mc Graw-Hill Book Company Inc., New York, p. 381.

WHO (1984a). *Guidelines for Drinking Water Quality: Recommendations*. World Health Organization, Geneva, 1: 130.

WHO (1984b). *Guidelines for Drinking Water Quality: Health Criteria and Other Supporting Information*. World Health Organization, Geneva, 2: 335.

WHO (1985). *Guideline for Drinking Water Quality: Drinking Water Quality Control in Small Community Supplies*, Geneva, 3: 1.

Chapter 5

Ecosystem Services of *Casuarina equisetifolia* for Sustainable Atmospheric CO_2 Build-Up in Tropical Farm Forestry

M. Uma[1], E. Mohan[1] and K. Rajendran[1]*

ABSTRACT

Trees are vital to fight against global warming, because they absorb and store the key greenhouse gases before they reach the atmosphere. While all living plants absorb CO_2 as part of photosynthesis, trees absorb more due to their size and extensive root structure. Hence, a forestation has been suggested as a cost effective approach to mitigate the predicted effects of global climatic change. Therefore, any increase in the forest biomass such as higher carbon recycling potential etc will reduce the build-up atmospheric carbon dioxide. There are number of forestry options available to stabilize the atmospheric CO_2 build-up. One of the most important option planting trees in agro-forestry and farm forestry. Casuarina (*Casuarina equisetifolia* Forst) is commonly used for wasteland development and suitable for farm forestry and agro-forestry plantation, due to

---

1 PG and Research Department of Botany, Thiagarajar College Kamarajar Salai, Madurai – 625 009, Tamil Nadu, India

* Corresponding Author E-mail: kuppurajendran@rediffmail.com

its ability to form symbiotic nitrogen fixing microorganism of Frankia and arbuscular mycorhizal association which is used for soil reclamation. Field experiment was conducted to study the growth, productivity and carbon accumulation of *Casuarina equisetifolia* Forst. in the farm forestry in east coast district of Villupuram in Tamilnadu. It is estimated that three years old C. *equisetifolia* was in the average height of 14.917 meter and 10.425 cm girth at breast height and basal area of 25.890 cm. It was estimated that the weight was 31.837 kg/tree above ground biomass, 5.255 kg/tree below ground and 37.092 kg/tree of total biomass. Totally 3,70,092 kg/ha of biomass and 1,48,827 kg/ha carbon was accumulated in three years oid *Casuarina equisetifolia* in tropical coastal region in India. In future carbon accumulation in *Casuarina* will find ways to efficiently manage the biosphere as sustainable carbon storage in farm forestry. Dissemination of innovative technology like Silviculture management, organic farming including use of biofertilizers, to the farmers, forest plantation corporation and tree growers to improve the growth, biomass and carbon pool of *Casuarina* is the viable option for the sustainable atmospheric CO_2 build-up in tropical plantation and agroforestry.

Keywords: *Biomass, Carbon accumulation, Casuarina equisetifolia, Farm forestry, Growth, and Productivity.*

Introduction

Trees promote sequestration of carbon into soil and plant biomass. Therefore, tree based land use practices could be viable alternatives to store atmospheric carbondioxide due to their cost effectiveness, high potential of carbon uptake and associated environmental as well as social benefit (Costa 1996).The tree sequester carbon in their tissues, and as the amount of tree biomass mitigate the carbon dioxide in atmosphere.

On the other hand, developing allometric regressions requires estimating biomass content and carbon of individual trees, a task that involves multiple steps. Furthermore, there is no universal standard for estimating the biomass or carbon of a tree. The general procedure for estimating biomass is to cut down a tree, weigh it, take samples of different tree components, and dry these components. The variation in carbon sequestrate capacity of plantations depended on DBH, height number of branches and crown canopy of individual plants. Increase in annual productivity of plantations directly indicates on increase in forest biomass and hence higher carbon sequestration potential. (Phanikumar *et al.*, 2009).

Casuarina equisetifolia is the fast growing tree in Coastal lands of Andhra Pradesh, Tamilnadu, Orissa, West Bengal, Maharashtra, Gujarat and Karnataka. It's also important species for the control of erosion, especially on Coasts. Its natural habitat and Sand dunes and on poor inland soils where it does well because of its ability to fix nitrogen. Thus its improving Soil fertility can contribute appreciable to carbon sequestration in Plantation soil and trees takeup CO_2 during photosynthesis and can store carbon in the roots and stems for long periods of time. The values of aboveground biomass and carbon sequestration in pure plantations are higher in *Casuarina equisetifolia*. (Parrotta, 1999; Silver *et al.*, 2000; Subak, 2000). Thus its improving Soil fertility can contribute appreciable to carbon sequestration in Plantation soil. Most farmers appear to be unaware of the growth of the voluntary carbon sector. Biomass Carbon sequestration also focused on the economic benefits of carbon trading by addition income that local farmers make from selling carbon credits. (Hamilton *et al.*, 2007).Therefore, it is necessary to estimate the growth, biomass and nutrient removal in the harvested biomass, which in turn, raises concerns about long-term site quality and sustainable production. Hence the present research work was aimed (a) to estimate the biomass allocation total biomass (b) to estimate the tissue nutrient concentration and carbon storage of Casuarina.

Materials and Methods

Study Area

The study was conducted in the farm land at Panjamadevi Villianur village in Villupuram district in South-Eastern portion of the state of Tamil Nadu, India is located at 11°59′–12° 48′ N latitude and 78° 60′–79° to 80° E longitude, at an elevation of 45 m above msl. (plate.1). The temperature of this study site ranges from 32 to 24°C and the annual precipitation averages between 1001 and 1200 mm. The soil type is sandy clay (sand 74.6 per cent: silt 10.4 per cent: clay 15 per cent) with a pH 8.6. Organic content of the soil was 1.58 per cent. The nitrogen content of soil was 98, phosphorus 4, potassium 88 kg/acre respectively and magnesium, copper, ferrous, zinc contents were 2.34, 0.60, 9.90, 0.60 and 0.88 ppm respectively.

Experimental Design and Planting

C. equisetifolia seedlings were obtained from Tamil Nadu Newsprint and Paper Limited (TNPL), Karur, Tamil Nadu, India. Six month old seedlings were transplanted in 30 cm^3 pits at a spacing of 1x1 m. The experiment was set up in a randomized block design (RBD). A total of 432 (36 plants in a block with 12 replicates) plants were used for data collection. After planting, the field was kept free from weeds through periodical hand weeding and hoeing once in two months during the study period. Watering was done twice a month during summer season.

Estimation of Soil Physico-Chemical Analysis

The soil (0–30, 30-60 and 60-90 cm) was sampled by a 45-mm-diameter hand auger with in a experimental plot. Visible roots and organic residues were removed at the sampling time, and then each sample soil was air-dried, and then was ground, and stored in plastic bags before analysis. (Qingkui Wang *et al.*, 2007). Soil pH was measured using a mixture of soil and deionized water (1:2.5, w/v) with a glass electrode (Jackson, 1973); total phosphorus (P) and available P were measured colorimetrically (Jackson, 1973); total K and available K by flame photometry were determined. Calcium and magnesium were determined with neutral normal ammonium acetate solution by Versenate method (Jackson, 1973). Organic carbon was estimated by Walkley and Black wet digestion method (Piper, 1966).

Biometri Observations

Estimation of Growth

The girth at breast height (1.37m) and the total height, basal area was measured with measuring tape. Basal area was used to calculate volume of the trees. The volume of individual trees was estimated following (Ravichandran *et al.*, 2003).

Estimation of Biomass

Random sampling method was adopted to estimate the biomass. Biomass components of tree (*i.e.*, bole wood, bark, branch, twig, leaf and root) were separated and fresh weight was estimated. Roots were excavated from 1 m^3 soil volume for three randomly chosen harvested trees in each year. 100 gram of fresh samples for different components (such as needle, branch, stem, bark and root) were brought to the laboratory for fresh weight was estimated, then plant samples were also oven dried at 80°C and taken for dry weight estimation and nutrient analysis.

Carbon Analysis

The oven-dried plant samples were ground to pass through a 0.5 millimetre plastic sieve before digestion for the biochemical analysis. Total C was measured using the dry combustion with a CHNS analyzer (VARIO-ELIII elementar).

Statistic Analysis

Statistical analysis were used to determine the statistical significance for differences in growth, above and below ground biomass and carbon sequestration were separated using SYSTAT 10 Software Package.

Results

Growth

This study present results of measurements of diameter, total height, volume, basal area, above and below ground biomass, and carbon sequestration potential of Casuarina plantation established in east coast district of Tamil Nadu, India. Growth rate of 3 years old *C.equisetifolia* in farm forestry plantation shows that the growth of the height was recorded in 14.91m. GBH and Basal area showed 10.42cm and 25.89cm respectively (Table 5.1). Figure 5.1 clearly showed linear relationship with significant correlation (r^2=0.08) of 3 year old Plantation trees.

Tree Biomass and Carbon Storage in Tree

Biomass variation of different components in different age groups of *Casuarina equisetifolia* exhibited in the Table 5.2. Above ground biomass allocation of three year old tree showed as in the stem (61 per cent) > Needle (21 per cent) > branch and twig (12 per cent) > bark (6 per cent). The below ground biomass in three year old trees recorded 99 per cent root biomass and 1 per cent root nodules biomass. Total biomass allocation showed as in the stem (47.5 per cent) > Needle (20.9 per cent) > root (13.2 per cent) > branch and twig (11 per cent) > bark (5.4 per cent) > fruit (1.1 per cent) and root nodules (1.0 per cent) (Figure 5.2).

Table 5.1: Growth and Biomass of Different Components in Three Years Old *Casuarina equisetifolia* in Coastal Region Farm Forestry

GBH (cm)	10.425±0.24
Height (m)	14.917±0.24
Basal Area (cm)	25.89±0.52
Stem weight (kg)	17.632±0.072
Bark weight (kg)	1.999±0.072
Branch and Twig Weight (kg)	4.072±0.089
Needle Weight (kg)	7.735±0.069
Fruit Weight (kg)	0.399±0.010
Total agb (kg)	31.837±1.16
Root Weight (kg)	4.895±0.120
Root Nodule (kg)	0.360±0.014
Total bgb (kg)	5.255±0.171
Total biomass (kg)	37.092±0.520

Table 5.2: Biomass of Different Components in Three Years Old *Casuarina equisetifolia* in Coastal Region Farm Forestry

Stem weight (kg)	7.479±0.324
Bark weight (kg)	0.598±0.016
Branch and Twig Weight (kg)	1.425±0.048
Needle Weight (kg)	3.035±0.132
Fruit Weight (kg)	0.146±0.005
Total agb (kg)	12.683±0.532
Root Weight (kg)	1.992±0.071
Root Nodule (kg)	0.207±0.08
Total bgb (kg)	2.199±0.103
Total biomass (kg)	14.882±0.485

Carbon storage in the tree components of *Casuarina* plantation is showed in Figure 5.1. It was considered above ground biomass carbon was recorded in 12.683 (Kg/tree). Whereas in the case of below ground biomass was recorded in 2.199 (Kg/tree). Totally 14.882 (Kg/tree) Carbon was accumulated in three year old trees of *C.equisetifolia* (Table 5.2).

Predicted Biomass and Biomass Carbon

In this plantation, regression model (DBH as Biomass) of biomass (Kg/tree) ranged from average biomass was 37.092 Kg/tree and presented r^2 values as 0.90 to 0.96 in three year old trees and r^2 values are significantly positive correlation ($P<0.05$) in this study. In Villupuram plantation, predicted average biomass carbon was 14.882 Kg/tree and presented r^2 values as 0.84 to 0.96 (Table 5.3). In 1-3 years old plantation

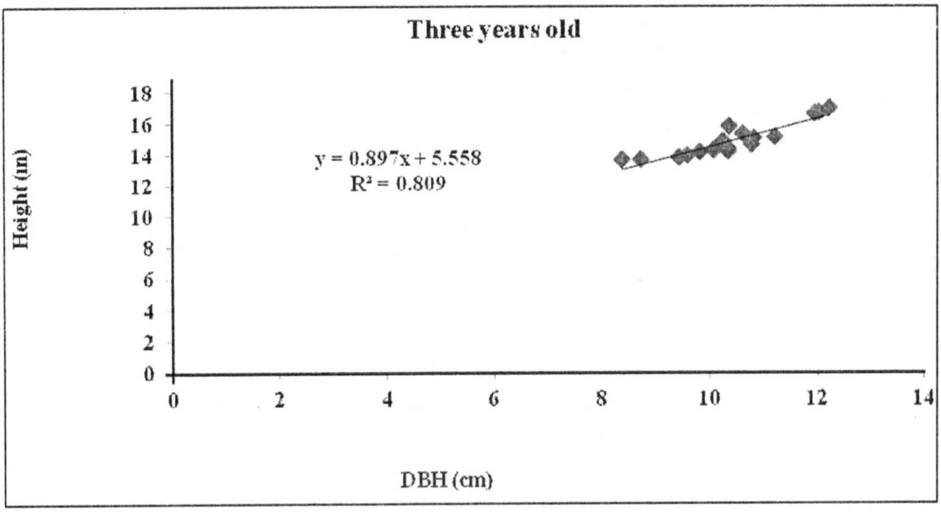

Figure 5.1: Relationship between Height and DBH of *Casuarina equisetifolia* Plantation

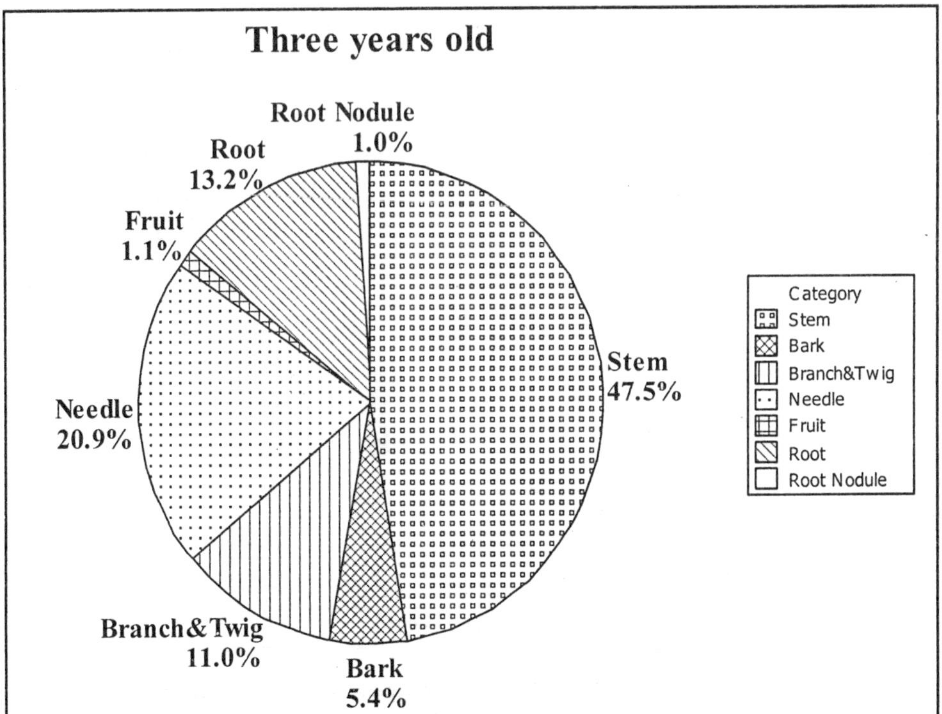

Figure 5.2: Relative per cent Contribution of Biomass of Different Tree Components to the Total Estimated Biomass in *Casuarina equisetifolia* Plantation

of this site, relationship between total biomass carbon and age of *Casuarina equisetifolia* plantation indicated positively correlation (P<0.05) (Figure 5.3).

Table 5.3: Allometric Models Used in the Present Study to Estimate Total Biomass and Carbon in Kilograms for *Casuarina equisetifolia* Plantation

Year of Plantation	*Tree Components*	*Regression Equitation*	*r²*
	Stem	ln(biomass)= 1.858+0.431 ln(DBH)	0.90
	Bark	ln(biomass)= -2.724+1.455 ln(DBH)	0.91
	Branch&Twig	ln(biomass)= -0.517+0.819 ln(DBH)	0.90
	Needle	ln(biomass)= 1.223+0.351 ln(DBH)	0.92
	Root	ln(biomass)= -0.704+0.977ln(DBH)	0.95
	Total Biomass	ln(biomass)= 2.260+0.574 ln(DBH)	0.96
Three year old			
	Stem	ln (C) = 0.739 +0.571 ln(DBH)	0.88
	Bark	ln (C) = -3.569+1.448 ln(DBH)	0.84
	Branch&Twig	ln (C) = -1.776 +1.003 ln(DBH)	0.83
	Needle	ln (C) = 0.341+0.351 ln(DBH)	0.92
	Root	ln (C) = -1.861 +1.165 ln(DBH)	0.91
	Total Biomass	ln (C) = 1.167 + 0.699 ln(DBH)	0.96

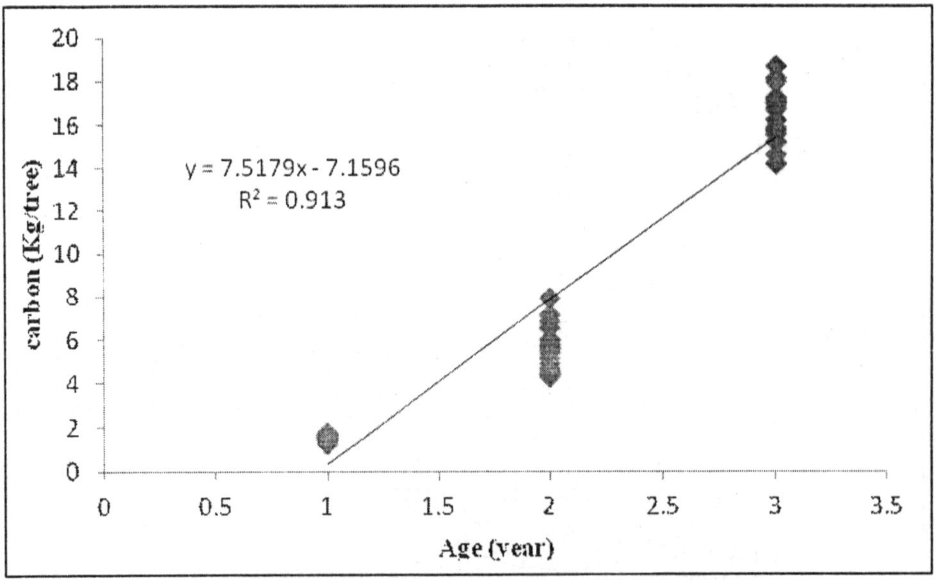

Figure 5.3: Relationship between Age and Carbon Content of *Casuarina equisetifolia* Plantation

Correlation

All the measurable parameters gave highly positive and significant correlation (Table 5.4).

Table 5.4: Correlation Values for Tree Growth, Biomass and Biomass Carbon of *Casuarina equisetifolia* Plantation at Farm Foresrtry

Characters	*DBH (cm)*	*Height (m)*	*Basal Diameter (cm)*	*Total Biomass (Kg)*
DBH (cm)	1.000			
Height (m)	0.88*	1.000		
Basal diamete (cm)	0.86**	0.900**	1.000	
Total Biomass (Kg)	0.78*	0.868*	0.901**	1.000

* Correlation is significant at the 0.05 level (2-tailed).

** Correlation is signicfiant at the 0.01 level (2-tailed).

Discussion

Casuarina Growth

The results of the present study provide valuable information to support the establishment of Casuarina plantations and knowledge exists regarding several aspects of this species growth, biomass, nutrient accumulation, litter production and nutrient cycling for site quality management. Casuarina growth depends on light, moisture and available nutrients. The excellent growth of *Casuarina equisetifolia* was estimated in the height of 14.917 m and GBH of 10.425 centimeter in three years old tree after planting and which comparable with the earlier study of Rajendran and Devaraj (2004) they were estimated in the height of *C. equisetifolia* ranges from 9.87–11.90m and GBH ranges from 16.8–23.2 centimeter in two years after planting inoculated with biofertilizers. In 3 year old plantations of dbh (cm) and height (m) recorded as 6.1 and 11.0 in Elverde respectively. Dbh (cm) and height (m) of Lajas recorded in 1 year old (4.4 and 3.7) and 2 year old (10.8 and 9.0) respectively (International Institute of Tropical Forestry).However, *C. equisetifolia* attained 204.8 cm height and 3.6 cm diameter after two years under arid conditions in Egypt (Lakany 1983) and incomparable with *Casuarina equisetifolia* grown in different parts of the world, the height and diameter of Casuarina 1, 2 and 3 years old shows 1.6, 2.2, and 2.5 m respectively in the Philippines (Halos 1983). Slow growth also found in recent estimate of *C. equisetifolia* trees (10 years old) were, height (11.4–27.9 m), girth (1.11–2.42 m) and basal area (0.098–0.466 m^2) Emmanuel Ehiabhi Ukpebor *et al.* (2010). In our present estimate the growth improvement may attribute that improved genetic material obtained from TNPL and systematic and scientific cultivation of Casuarina.

Biomass

According to these results, it seems that fast-growing species (*i.e. Casuarina equisetifolia*) accumulate biomass and carbon very fast in the initial stage of their lifespan, before they are about 3 years in the farm forestry condition in well managed plantation. Our estimation of above and below ground biomass allocation of 3 years old *Casuarina equisetifolia* was 37.092 kg/tree. Casuarina planted in similar agroclimatic condition in Tamil Nadu (Cuddalore) were regarded above ground biomass of 39.9 MT ha^{-1} $year^{-1}$ in Alfisols and 15.5 MT ha^{-1} $year^{-1}$ in Vertisols (Ravi *et al.*, 2010). In East Timor, *C. equisetifolia* (=30 years) contain up to 200 $MgCha^{-1}$ in above ground biomass carbon (Lasco and Cardinoza, 2007).

Biomass Carbon

Trees promote sequestration of carbon into soil and plant biomass. In our estimates 14.882 kg/tree of carbon stored in three year old *Casuarina equisetifolia* which comparable with Casuarina planted in Puerto Rico by Parrotta (1999) estimated Carbon storage (Mg/ha/yr) in four-year-old plantation of *Casuarina equisetifolia*. The values of aboveground biomass and carbon sequestration in pure plantings are higher than values found in other regions of tropical humid climate such as in 8.5- year-old pure plantings of *Casuarina equisetifolia*, *Eucalyptus robusta*, and *Leucaena leucocephala* in Puerto Rico (Parrotta, 1999). Values of this study are also higher than those reported for pure plantation of *Pinus caribaea*, *Leucaena* spp., *Casuarina* spp., *Pinus patula*, *Cupressus lusitanica*, *Acacia nilotica*, *Senna siamea*, *Azadirachta indica*, *Gmelina arborea* (Schroeder, 1992; Brown *et al.*, 1986, Lugo and Brown, 1992; Silver *et al.*, 2000; Subak, 2000).

Conclusion

Casuarina equisetifolia in high density plantation in farm forestry systems have a higher potential to produce more biomass and sequester carbon than pastures, or field crops, because tree incorporation in croplands and pastures would result in greater net aboveground as well as belowground carbon sequestration in short rotation. At the same time *C. equisetifolia* trees harvesting of short rotations (*e.g.* 3 years of age) will lead to significant removals of available nutrients, especially the mobile nutrients, as this is the period of age when nutrients are in high concentrations in the tree components. Total nutrient replacement, taking other loss factors also into account, will need to be factored into the costs of plantation systems in order to maintain long-term productivity and maintain sustainable systems. That is, in long-term conservative rotations, part of the nutrient capital of the site is harvested but in quantities which are not significant, whereas short rotations need to be considered as a farming system and nutrients need to be supplied because the harvesting regime will potentially remove a significant component of the nutrient capital of the site.

Acknowledgements

This work was supported by the Jawaharlal Nehru Memorial Fund, New Delhi. (Doctoral fellowship for Ms.M.Uma, Research scholar). I thank many individuals who provided assistance in the field and laboratory work.

References

Brown, S., Lugo, A.E. and Chapman, J. (1986). Biomass of tropical plantations and its implications for the global carbon budget. *Canadian Journal of Forest Research*. **16 (2)**: 390–394.

Costa, P.M. (1996). Tropical forestry practices for carbon sequestration–review and case study from south East Asia. *Ambia*. **25**: 279-283.

Emmanuel Ehiabhi Ukpebor, Justina Ebehirieme Ukpebor, Emmanuel Aigbokhan, Idris Goji, Alex Okiemute Onojeghuo and Anthony Chinedum Okonkwo1. (2010). *Delonix regia* and *Casuarina equisetifolia* as passive biomonitors and as bioaccumulators of atmospheric trace metals *Journal of Environmental Sciences.* **22(7)**: 1073–1079.

Halos, S.C. (1983). Casuarina in Philippine forest development. *In*: Midgley, S.J., Turnbull, J.W. and Johnston, R.D. (Eds.), Casuarina Ecology management and Utilization. *Proceedings of First International Casuarina Workshop, CSIRO, Melbourne, Australia.* pp. 89-98.

Hamilton, K., R. Bayon, G. Turner and G. Higgins. (2007). "State of the voluntary carbon markets 2007: Picking up steam" *London, New Carbon Finance and Washington, D.C., Ecosystem Marketplace.*

Jackson, M.L. (1973). Soil chemical analysis. Printice hall of India (Pvt) Ltd., New Delhi.

Lakany, M.H.E.L. (1983). A review of breeding drought resistant Casuarina for shelter belt establishment in arid regions with special reference to Egypt. *Forest Ecology Management*, **6**: 129-137.

Lasco, R.D., Cardinoza., M.M. (2007). Projection of future carbon benefits of a carbon sequestration project in east timor. *Mitigation and Adaptation Strategies for Global Change.* **12**: 243–257.

Lugo, A., Wang, D. and Bormann, H.A. (1990). Comparative analysis of biomass production in five tree species. *Forest Ecology and Management.* **31**: 153–66.

Parrotta, J.A. (1999). Productivity, nutrient cycling, and succession in single- and mixed-species plantations of *Casuarina equisetifolia, Eucalyptus robusta,* and *Leucaena leucocephala* in Puerto Rico. *Forest Ecology and Management* **124**: 45-77.

Phanikumar, C., Ashil Guutosh.A., Murkute., Sunil Gupta and Shashi Bala Singh. (2009). Carbon sequestration with special reference to agroforestry in cold deserts of Ladakh. *Current Science*, **97(7)**: 1063-1068.

Piper, C.S. (1996). *Soil and Plant Analysis.* Hans Publications, Bombay.

Rajendran, K. and Devaraj, P. (2004). Biomass and nutrient distribution and their return of *Casuarina equisetifolia* inoculated with biofertilizers in the farm land. *Biomass and Bioenergy*, **26(3)**: 235-249

Ravi, R., Buvaneswaran., C.S and Vijay Kumar (2010). Financial and carbon sequestration potential of *casuarina equisetifolia* plantations: A case study in Cuddalore district of Tamil Nadu. *International Congress of Environmental Research.* (ICER-10).

Ravichandran V.K. and T.N. Balasubramaniam and M. Jeyachandran (2003). Development of Volume table for *Casuarina equisetifolia. Indian Journal of Forestry.* **26(1)**: 7-10.

Schroeder, P. (1992). Carbon storage potential of short rotation tropical tree plantations. *Forest Ecology and Management*. **50**: 31–41.

Silver, W.L., Ostertag, R. and Lugo, A.E. (2000). The potential for carbon sequestration through reforestation of abandoned tropical agricultural and pasture lands. Restor. *Ecology*. **8(4)**: 394–407.

Subak, S. (2000). Forest protection and reforestation in Costa Rica: Evaluation of a clean development mechanism prototype. *Environment Management*, **26(3)**: 283–297.

Chapter 6

Optimization Doses of Different Bioagents Against *Fusarium* Wilt of Pigeonpea

D.N. Dhutraj[1]*, P.R. Bhandarge*[1]*,*
M.G. Korde[1] *and A.L. Harde*[1]

ABSTRACT

Pigeonpea wilt is caused by *Fusarium udum* is becoming major threat in the cultivation of the crop. Management through chemicals adds toxins in the ecosystem and due to its non judicious use, cost and soil population alternative means of disease management is envisaged. *Trichoderma viride, Trichoderma harzianum* and *Pseudomonas fluorscense* are the important biological agents for disease management was evaluated individually with different doses under artificial epiphytotic conditions. Experiment was conducted at college of Agriculture, Latur in *Kharif* 2009 on susceptible Cv. ICP-2376 to asses the effect of seed dressing doses of *T. viride, T. harzianum* and *P. fluorescens* in management of wilt disease and maximum efficacy was observed with 30 g/kg dose of antagonists followed by 25 g/kg. Per cent decrease of *Fusarium* wilt varied from 73.42 per cent and 65.92 per cent indifferent doses of antagonist over control (90.00 per cent). Maximum wilt decrease being in 30 g/kg dose at 100 day of crop growth. Maximum seed germination, vigour index and dry matter was observed in 30 g/kg dose of antagonist followed by 25 g/kg.

Keywords: Bioagents, Doses, Fusarium wilt, Pigeonpea.

---

1 College of Agriculture, Latur – 413 512, M.S., India

Introduction

Among the different pulses grown in India, pigeonpea [*Cajanus cajan* (L.) Millsp.] is the most important kharif pulse crop with relatively much higher productivity. In India, pigeonpea is mainly known as red gram, arhar and tur. It is an important source of protein in the cereal based vegetarian diets. The protein content in different cultivars ranges from 15.5 per cent to 28.8 per cent. It is largely consumed in the form of split pulse or dal. In common with other legumes, pigeonpea has always been a very important component of the farming system in India because of its ability to fix atmospheric N_2. Being a deep rooted crop, it can thrive well under rainfed conditions, and improve soil structure.

In Maharashtra, the area under pigeonpea was 11.59 Lakh ha, production and productivity was 10.76 million tones and 928 kg/ha respectively. In Marathwada, area under pigeonpea was 4.67 lakh ha, production and productivity was 4.42 lakh tones and 1789 kg/ha respectively. In Latur district, area under pigeonpea was 3251 ha, production 3364 tonnes and productivity 1035 kg/ha respectively (Anonymous 2007-2008).

Pigeonpea wilt is caused by *Fusarium udum* is becoming major threat in the cultivation of the crop. Management through chemicals adds toxins in the ecosystem and due to its non judicious use, cost and soil population alternative means of disease management is envisaged. *Trichoderma viride, Trichoderma harzianum* and *Pseudomonas fluorscense* are the important biological agents for disease management was evaluated individually with different doses under artificial epiphytotic conditions.

Material and Methods

The pot culture experiment was conducted in sick soil. For this, the culture of *F. udum* was multiplied on Sand: Maize flour medium (1:1). 15 g maize flour and 85g of River bed sand are mixed thoroughly and filled in the conical flask of 250 ml capacity (50 g/flask) and sterilized in autoclave at 15lbs for 30 minutes. Then these flasks were inoculated aseptically with pure culture of *F. udum* and incubated at room temperature for 15 days. After 15 days of incubation, the inoculum was taken out from the flask and mixed with thoroughly with sterilized sand + soil (1:1) @ 100g inoculum per kg soil. This mixture (sand + soil + inoculum) was filled in earthen pots (25 cm dia.) sterilized with 5 per cent solution of copper sulphate, watered periodically and incubated for a week for colonization of the test fungus in pot soil. Thus the soil was made sick by inoculating inoculum several times. The seeds of highly susceptible variety ICP-2376 were treated separately with *T. viride, T. harzianum* and *P. fluorescens* @ 4g, 8g, 10g, 15g, 20g, 25g and 30g per kg seed respectively. Then treated seeds were sown @ 10 seeds/pot. The earthen pot with uninoculated soil served as control. All these pots were then watered regularly and kept in glass house for further observations. The observations was recorded after 100 days of crop growth on germination percentage, wilt incidence and vigour index and calculated by using Vincent (1947) formula

$$\text{Per cent disease control (PDI)} = \frac{C-T}{C} \times 100$$

where,

C: Per cent wilt incidence in the control

T: Per cent wilt incidence in the treatment

Vigour index = Root length (cm) + Shoot length (cm) x per cent germination

Results and Discussion

Optimization Doses of *Trichoderma viride*

The data presented in Table 6.1, revealed that doses of *T. viride* influenced significantly in management of wilt disease of Pigeonpea from one month crop growth onwards. Maximum efficacy was observed with 30 g/kg dose of antagonist followed 25 g/kg. Commonly used 4 g/kg dose of antagonist was inferior to other higher doses. Per cent decrease in wilt over control varied from 40.64 to 68.79 per cent in different doses of antagonist, maximum decrease being in 30 g/kg doses followed 25 g/kg dose. Maximum stimulatory effect was observed on germination percentage (93.33 per cent), mean per cent wilt (16.67 per cent), per cent wilt control (68.79 per cent), vigour index (2926) and dry matter (Average of ten plants) (32.76 g) at 30 g/kg dose followed by 25 g/kg dose of the antagonist.

Optimization Doses of *Trichoderma harzianum*

All the doses of *T.harzianum* bioagent reduced per cent wilt from 23.33 to 50.00 per cent significantly as compared to check. Maximum reduction in mean disease wilt percentage (23.33 per cent) was recorded with *T. harzianum* @ 30 g/kg seeds followed by 25 g/kg seeds (26.66 per cent). Maximum per cent wilt control was observed in 30 g/kg dose (73.42 per cent) followed by 70.09 per cent at 25 g/kg dose. It was also observed that maximum vigour index (3313) and dry matter (34.26 g) at 30 g/kg dose followed by 25 g/kg dose in connection with vigour index (3093) and dry matter (33.13 g) (Table 6.1).

Optimization Doses of *Pseudomonas fluorescens*

Table 6.1 indicated that maximum mean per cent wilt (16.67), per cent wilt control (65.92), vigour index (3045) and dry matter (32.63 g) was significantly observed @ 30 g/kg seed dose of *Pseudomonas fluorescens* against *Fusarium* wilt of pigenpea. The next best treatment was observed @ 25 g/kg seed in connection with mean per cent wilt (26.67 per cent), per cent wilt control (58.14 per cent), vigour index (2703) and dry matter (30.03 g) with the influence of *Pseudomonas fluorescens.*

This experiment was aimed to find out the optimum dose of *T. viride, T. harzianum* and *P. fluroscens* to manage the *Fusarium* wilt of pigeonpea. The dose of *T. viride, T. harzianum* and *P. fluroscens* @ 30 g/kg seed was found significantly superior over the rest of doses. These preliminary studies have shown some promise that application of bioagents to achieved effective disease management lies between 25 to 30 g/kg seed. Present investigations are in conformity with, Kurundkar and Bombale (2008) conducted pot culture experiment in *F. udum* inoculated soil on pigeonpea var. BDN-2 to assess the effect of seed dressing doses of *T. harzianum* and reported that higher dose of *T. harzianum* were superior to lower dose in management of wilt disease.

Table 6.1: Optimization Doses of Different Bioagents Against *Fusarium* Wilt of Pigeonpea

Bioagents Doses	*Trichoderma viride*					*Trichoderma harzianum*					*Pseudomonas fluroscense*				
	Seed Germination (%)	*Mean % Wilt*	*% Wilt Control*	*Vigour Index*	*Dry Matter (g)*	*Seed Germination (%)*	*Mean % Wilt*	*% Wilt Control*	*Vigour Index*	*Dry Matter (g) (%)*	*Seed Germination*	*Mean % Wilt*	*% Wilt Control*	*Vigour Index*	*Dry Matter (g)*
4 g/kg	76.67* (50.22)	46.67 (27.85)	40.64	2043	26.65	80.00* (53.89)	50.00 (30.14)	43.61	2165	27.16	73.33* (47.32)	50.00 (30.14)	39.25	1921	26.20
8 g/kg	80.00 (53.89)	43.33 (25.71)	44.81	2144	26.90	80.00 (53.12)	46.66 (28.00)	47.77	2205	27.56	76.67 (50.22)	53.33 (32.28)	42.59	1998	26.06
10 g/kg	83.33 (56.79)	43.33 (25.71)	51.85	2354	28.20	86.67 (60.47)	40.00 (23.67)	54.72	2432	28.03	80.00 (53.89)	50.00 (29.99)	46.29	2168	27.10
15 g/kg	86.67 (60.47)	36.67 (21.63)	59.72	2455	28.33	83.33 (56.79)	36.67 (21.53)	58.59	2443	29.30	83.33 (56.79)	46.67 (27.85)	49.99	2291	27.50
20 g/kg	83.33 (57.57)	30.00 (17.52)	60.55	2441	29.90	90.00 (69.08)	30.00 (17.52)	66.01	2644	29.36	86.67 (60.47)	36.67 (21.53)	54.07	2479	28.70
25 g/kg	93.33 (72.76)	26.67 (15.54)	64.25	2873	30.76	93.33 (72.76)	26.66 (15.48)	70.09	3093	33.13	90.00 (69.08)	26.67 (14.48)	58.14	2703	30.03
30 g/kg	93.33 (72.76)	16.67 (9.64)	68.79	2926	32.76	96.67 (81.37)	23.33 (13.50)	73.42	3313	34.26	93.33 (72.76)	16.67 (9.60)	65.92	3045	32.63
Control	70.00 (44.80)	90.00 (69.08)	00	1396	20.00	70.00 (44.42)	90.00 (69.08)	00	1400	20.00	70.00 (44.80)	93.33 (72.76)	00	1397.67	20.00
SE ±	5.64	4.02	5.40	117.6	0.6	6.69	5.2	3.88	130.41	0.4	5.76	3.69	6.93	117.74	0.5
CD at 5%	17.10	12.18	16.35	356.18	1.86	20.28	15.75	11.76	394.97	1.253	17.45	11.17	21.00	356.61	1.64

Figures in parenthesis are angular transformed values.

Maximum efficacy was observed with 30 g/kg dose of the antagonist followed by 25 g/kg commonly used 4 g/kg dose of antagonist was inferior to other higher doses.

The optimization of doses of bioagents in pigeonpea and other crops were attempted earlier by workers (Prasad *et al.*, 2002; Gholve and Kurundkar, 2002; Pandey and Upadhyaya, 1999 and Khan and Sinha, 2005).

References

Anonymous (2007-2008). Area, production and productivity of pigeonpea. *Annual Report of AICPIP*, Kanpur.

Gholve, V.M. and Kurundkar, B.P. (2002). Biological control of pigeonpea wilt by *Pseudomonas fluorescens* and *Trichoderma viride. Indian J. Pulses Res.*, **15(2)**: 174-176.

Khan and Sinha (2005). Influence of different factors on the effectivity of fungal bioagents to manage rice sheath blight in nursery. *Indian Phytopath.*, **58**: 289-293.

Kurundkar, B.P. and Bombale, K.M. (2008). Effect of inoculum load of *Trichoderma harzianum* in management of pigeonpea wilt. *Indian Phytopath.*, **61(3)**: 416.

Pandey, K.K. and Upadhyay, J.C. (1999). Comparative study of chemical, biological and integrated approach for management of *Fusarium* wilt pathogen *J. Mycol. Pl. Pathol.*, **29(2)**: 214–216.

Prasad, R.D., Rangeshwaran, R., Hegde, S.V. and Anuroop, C.P. (2002). *Crop Protection*, **21(4)**: 293-297.

Vincent, J.M. (1947). Distortion of fungal hyphae in the presence of certain inhibitors. *Nature*, **159**: 85.

Chapter 7

In vitro Evaluation of Fungicides Against *Stemphylium vesicarium*, Causing Leaf Blight of Onion

A.P. Suryawanshi[1], K.T. Apet[1], S.H. Khade[1], D.P. Kuldhar[1] and A.L. Harde[1]

ABSTRACT

Foliage/leaf blight caused by *Stemphylium vesicarium* (Wallr.) E. Simmons is one of the major destructive disease of onion (*Allium cepa* L.) crop grown in the sate of Maharashtra. In the present study, bio- efficacy of eight fungicides was evaluated *in vitro* against *S. vesicarium*. All the fungicides tested were found fungistatic/fungicidal against the test pathogen and significantly inhibited mycelial growth of the test pathpgen over untreated control. However, Mancozeb 75 WP recorded significantly highest mean inhibition (90.01 per cent) of the test pathogen. The second and third best fungicides found were Carbendazim 50 WP and Copper oxychloride 50 WP which recorded mean growth inhibition, respectively of 89.25 and 86.86 per cent,. This was followed by the fungicides, *viz.*, Chlorothalonil 75 WP (84.77 per cent inhibition), Difenconazole 25 EC (84.02 per cent inhibition), Thiophanate methyl 70 WP (78.21 per cent inhibition), Penconazole 10 EC (77.61 per cent inhibition) and Hexaconazole 5 EC (76.43 per cent inhibition). Thus, in

---

1 Department of Plant Pathology, College of Agriculture, M.K.V., Parbhani – 431 402, M.S., India

the present study those fungicides found effective could be exploited under field conditions for effective management of *Stemphylium* leaf blight of onion.

Keywords: *Stemphylium vesicarium, Allium cepa, Fungicides, Inhibition.*

Introduction

Onion (*Allium cepa* L.) a bulbus spice (originated from Middle East Asia and introduced in India from Palestin) biennial herb is one of the most important vegetable crop grown throught world including India. Area, production and productivity of onion in India were, respectively 579.3 thousand ha., 7158.4 million tones and 12,357 kg/ha; and in the state of Maharashtra were 109.0 thousand ha., 1112.0 million tones and 10202 kg/ha. (Anonymous, 2008). In Maharashtra, the major districts growing onion are: Nashik, Pune, Dhule, Ahmednagar, Jalgaon, Aurangabad, Satara and Buldhana. The Onion crop is attacked by 66 diseases, of which 10 are bacterial, 38 fungal, six nematode, and three viral. Of these, two fungal diseases *viz.*, Purple blotch [*Alternaria porri* (Ellis)] and *Stemphylium* leaf blight (*Stemphylium vesicarium* (Wallr.) E. Simmons) are the most destructive diseases causing heavy yield losses. During recent years both the fungal diseases were found to occur in severe proportions under favorable environmental conditions particularly during *Kharif* and *Rabi* season. Foliage blight caused by *S. vesicarium* has been reported to inflict heavy yield losses to the tune of 80-90 per cent (Gupta, 1986; Shahanaz *et al.*, 2007) in onion crop.

Typical symptoms (Plate 7.1A) of the disease appeared on foliage and foliage sheath are: small, water soaked lesions of light yellow to brown coloured. As the lesions expand, they coalesce and cause extensive blightening of the leaves. Typically lesions are found in higher number on the sides of leaves facing wind. The centers of lesions turn brown, due to abundant sporulation of the pathogen, sometimes fruiting bodies called 'perithecia' may appear in infected tissues as small, black, pinhead-like raised bodies.

Considering, economic importance of the onion and destructive nature of foliage blight incited by *Stemphylium vesicarium* in onion, present studies on *in vitro* evaluation of fungicides against the test pathogen were undertaken during *Kharif* and *Rabi*, 2009 at the Department of Plant Pathology, College of Agriculture, Latur.

Materials and Methods

The bio-efficacy of eight fungicides *viz.*, Carbendazim 50 WP, Chlorothalonil 75 WP, Copper oxychloride 50 WP, Difenconazole 25 EC, Hexaconazole 5 EC, Mancozeb 75 WP, Penconazole 10 EC, and Thiophanate methyl 70 WP was evaluated (@ 500 and 1000 ppm each) *in vitro* against *S. vesicarium*, applying Poisoned food technique (Nene and Thapliyal, 1993). Requisite quantity of each test fungicide based on active ingredient was calculated and mixed thoroughly with autoclaved and cooled (40°C) Potato dextrose agar medium (PDA) in conical flasks to obtain desired concentrations of 500 and 1000 ppm. Fungicide amended PDA medium was then poured (5 Petri plates/fungicide/replication) aseptically in Petri plates (90 mm dia.). After solidification of the medium, all the plates were inoculated aseptically with 5 mm culture disc of the test fungus obtained from a week old actively growing pure culture of *S. vesicarium* (Plate 7.1B). Petri plate containing plain PDA (without fungicide) was inoculated with culture disc of the test pathogen, which served as untreated control. All the plates were incubated at 24 ± 1°C. The experiment was designed in CRD and all the treatments were replicated thrice.

When untreated control plate was fully covered with mycelial growth of the test fungus, observations on radial mycelial growth were recorded in all the treatment plates. Per cent inhibition of mycelial growth in treated plates was calculated by applying the formula given by Vincent (1927).

Table 7.1: *In vitro* Bio-efficacy of Fungicides at Different Concentrations on Radial Mycelial Growth and Inhibition of *S. vesicarium*

Fungicides	*MCD (mm)**		*Mean (mm)*	*% Inhibition*		*Mean %*
	500 ppm	*1000 ppm*		*500 ppm*	*1000 ppm*	
Thiophanate methyl	19.63	17.46	18.54	76.94 (61.30)	79.49 (63.07)	78.21 (62.17)
Mancozeb	9.50	7.50	8.50	88.84 (70.48)	91.19 (72.73)	90.01 (71.57)
Carbendazim	10.40	7.90	9.15	87.78 (69.53)	90.72 (72.26)	89.25 (70.86)
Copper oxychloride	12.63	9.73	11.18	85.16 (67.34)	88.57 (70.23)	86.86 (68.74)
Chlorothalonil	14.50	11.43	12.96	82.97 (65.62)	86.57 (68.50)	84.77 (67.02)
Penconazole	20.63	17.50	19.06	75.77 (60.57)	79.45 (63.04)	77.61 (61.75)
Hexaconazole	21.50	18.63	20.06	74.75 (59.83)	78.12 (62.11)	76.43 (60.95)
Difenconazole	14.70	12.50	13.60	82.73 (65.44)	85.32 (64.47)	84.02 (66.43)
Control	85.16	85.16	85.16	00.00 (00.00)	00.00 (00.00)	00.00 (00.00)
S.E. ±	0.25	0.16	–	0.10	0.09	–
C.D. (P=0.05)	0.76	0.49	–	0.31	0.29	–

*: Average of three replications.

Figures in parenthesis are angular transformed values.

MCD: Mean Colony Diameter (mm)

Results and Discussion

Results (Table 7.1, Plate 7.1, Figure 7.1) revealed that all the fungicides tested (@ 500 and 1000 ppm each) recorded a wide range of mean colony diameter (MCD)/ mycelial growth of the test fungus and it was ranged from 9.50 mm (Mancozeb) to

Plate 7.1A: Typical Symptoms of *Stemphylium* Blight on Onion

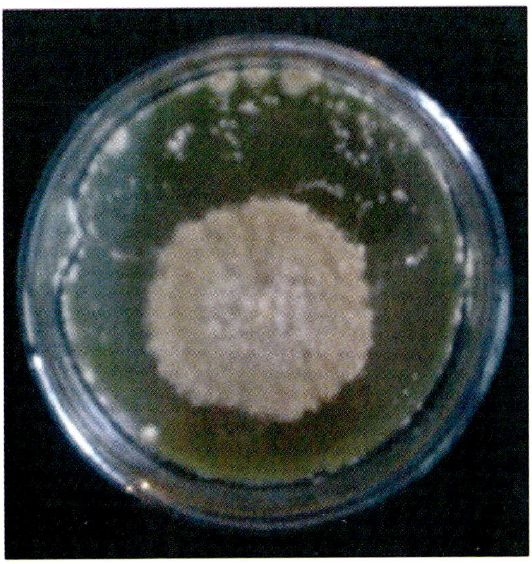

Plate 7.1B: Pure Culture of *Stemphylium vesicarium*

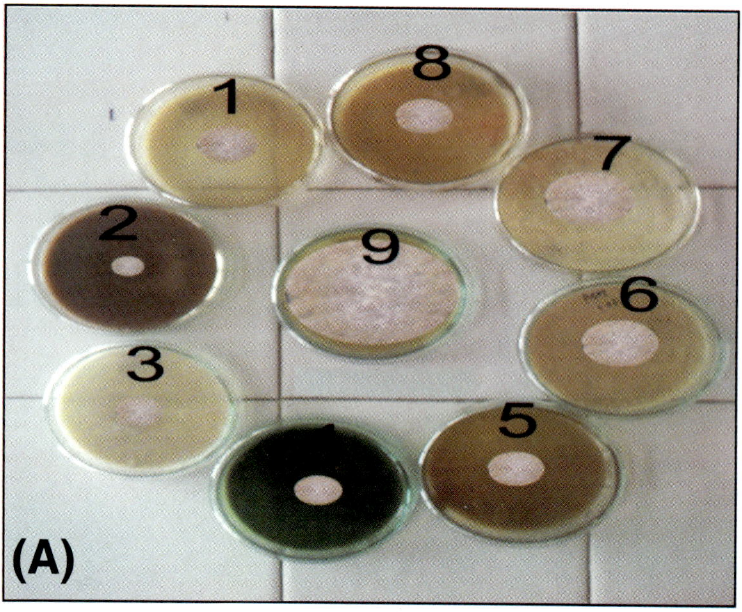

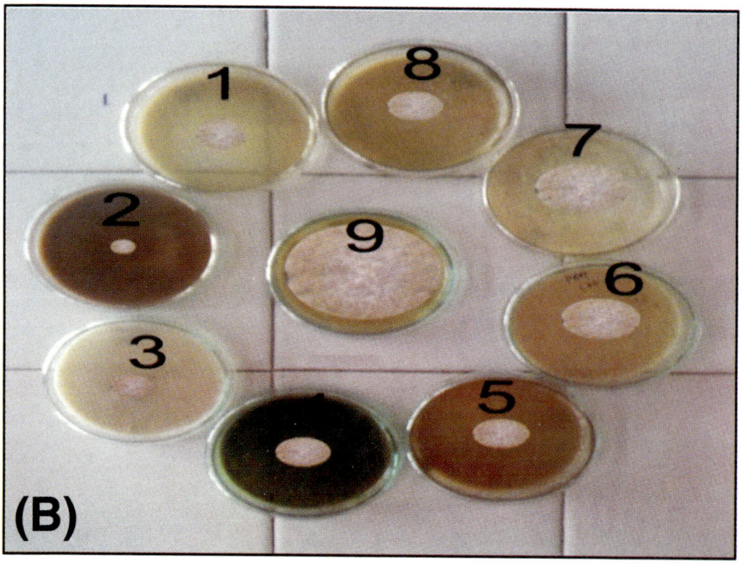

Plate 7.2: ***In vitro*** **Effect of Fungicides at 500 ppm (A), and 1000 ppm (B) on Mycelial Growth and Inhibition of** ***S. vesicarium***

1: Thiophanate methyl; 2: Mancozeb; 3: Carbendazim; 4: Copper oxychloride; 5: Chlorothalonil; 6: Penconazole; 7: Hexaconazole; 8: Difenconazole; 9: Control

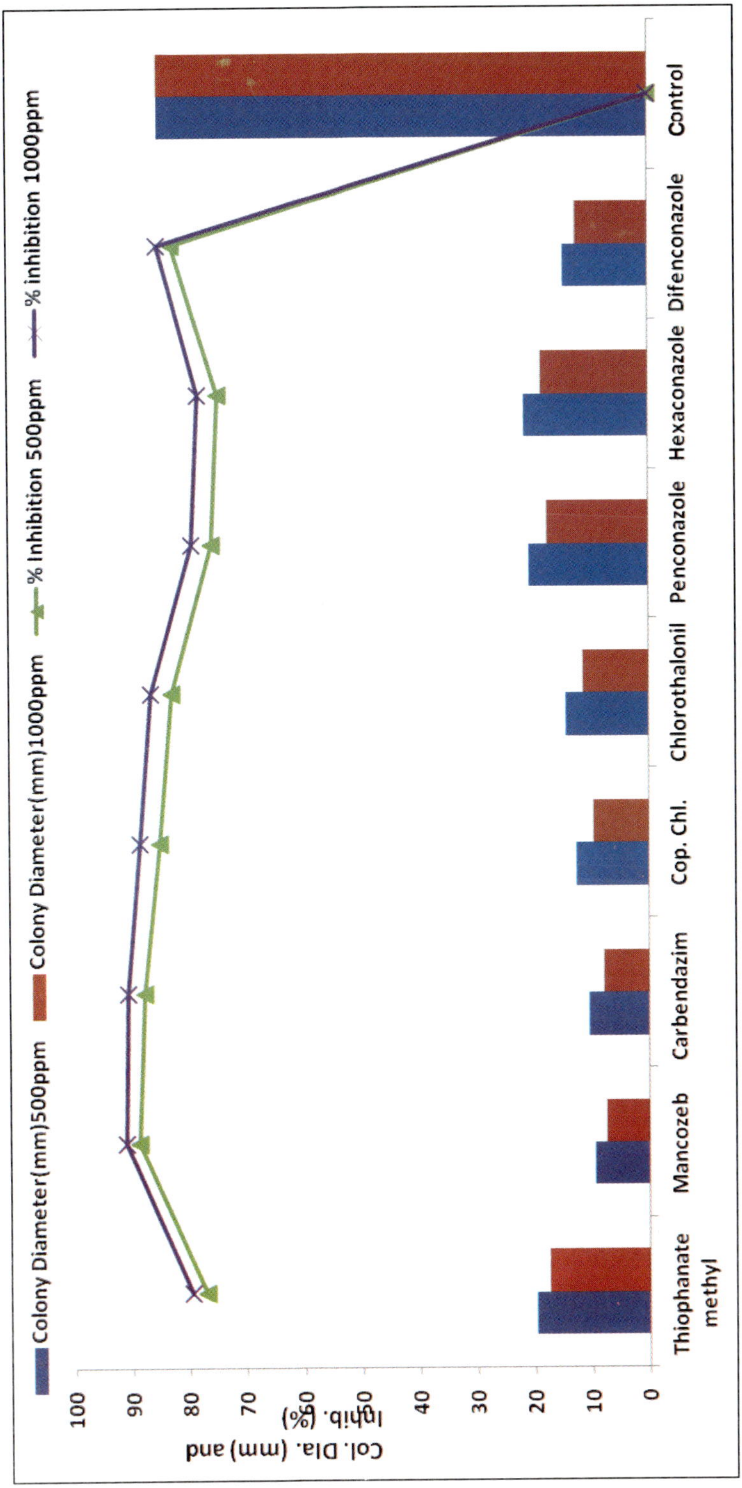

Figure 7.1: *In vitro* Bio-efficacy of Fungicides at Dfferent Concentrations on Radial Mycelial Growth and Inhibition of *S. vesicarium*

21.50 mm (Hexaconazole) at 500 ppm and 7.50 mm (Mancozeb) to 18.63 mm (Hexaconazole) at 1000 ppm. All the test fungicides significantly inhibited mycelial growth of the test fungus over untreated control (00.00 per cent). Further, it was found that percentage mycelial inhibition of the test pathogen was increased with increase in concentration of the fungicides tested (Plate 7.2).

At 500 ppm, percentage mycelial growth inhibition (Plate 7.2A) was ranged from 74.75 (Hexaconazole) to 88.84 (Mancozeb). However, highest percentage of mycelial growth inhibition was recorded with Mancozeb (88.84 per cent). This was followed by the fungicides, *viz.*, Carbendazim (87.78 per cent), Copper oxychloride (85.16 per cent), Chlorothalonil (82.97 per cent) and Difenconazole (82.73 per cent). Comparatively minimum mycelial growth inhibition was recorded with Hexaconazole (74.75 per cent) which was followed by Penconazole (75.77 per cent), Thiophanate methyl (76.94 per cent)

At 1000 ppm, similar trend of mycelial growth inhibition (Plate 7.2B) with the test fungicides was recorded and it was ranged from 78.12 per cent (Hexaconazole) to 91.19 per cent (Mancozeb). However, highest mycelial growth inhibition was recorded with Mancozeb (91.19 per cent). This was followed by the fungicides *viz.*, Carbendazim (90.72 per cent), Copper oxychloride (88.57 per cent), Chlorothalonil (86.57 per cent), Difenconazole (85.32 per cent), Thiophanate methyl (79.49 per cent), Penconazole (79.45 per cent) and Hexaconazole (78.12 per cent). Mean mycelial growth inhibition recorded with all the fungicides tested was ranged from 76.43 per cent (Hexaconazole) to 90.01 per cent (Mancozeb). However, Mancozeb was found most fungi–static and recorded significantly highest mean mycelial growth inhibition of 90.01 per cent. The second and third best fungicides found were Carbendazim (89.25 per cent), and Copper oxychloride (86.86 per cent). This was followed by Chlorothalonil (84.77 per cent), Difenconazole (84.02 per cent), Thiophanate methyl (78.21 per cent), Penconazole (77.61 per cent) and Hexaconazole (76.43 per cent).Thus, all the fungicides tested were found fungistatic against *S. vesicarium* and significantly inhibited mycelial growth of the test pathogen over untreated control. However, Mancozeb recorded highest mean mycelial growth inhibition (90.01 per cent), followed by Carbendazim (89.25 per cent), Copper oxychloride (86.86 per cent), Chlorothalonil (84.77 per cent), Difenconazole (84.02 per cent), Thiophanate methyl (78.21 per cent), Penconazole (77.61 per cent) and Hexaconazole (76.43 per cent).

Results (Table 7.1, Figure 7.1) revealed that, all the eight fungicides evaluated (@ 500 and 1000 ppm each) *in vitro* exhibited fungi–static reactions against *S. vesicarium*, causing leaf blight in onion; and significantly inhibited the mycelial growth (Plates 7.2A and B) over untreated control. The percentage mycelial inhibition of the test pathogen was found to be increased with increase in the concentrations of the test fungicides.

However, Mancozeb was found most effective and recorded significantly highest mean mycelial inhibition (90.01 per cent). The second and third best fungicides found were Carbendazim (89.25 per cent) and Copper oxychloride (86.86 per cent). Rest of the fungicides also recorded significant inhibition of the test pathogen which was ranged from 76.43 (Hexaconazole) to 84.77 (Chlorothalonil) per cent. Similar fungistatic effects of these fungicides against *S. vesicarium*, infecting onion and garlic and *Alternaria* spp. infecting many other crops were reported earlier by several workers. Fungicides Mancozeb, Carbendazim, Copper oxychloride, Chlorothalonil and Difenconazole were reported effective against *S. vesicarium* and *Alternaria* spp. earlier by several workers (Gupta and Srivastava, 1988; Barnwal, *et al.*, 2003; Amresh and Nargund, 2004; Kumari *et al.*,, 2006; Kharakwal *et al.*, 2006; Khosla *et al.*, 2007).

Thus, from the ongoing results and discussion it can be concluded that for effective management of the pathogen *Stemphylium vesicarium*, inciting foliage blight in onion; the fungicides *viz.*, Mancozeb, Carbendazim, Difenconazole, Copper oxychloride, Hexaconazole and Chlorothalonil could be exploited on large scale under field conditions.

References

Amaresh, Y.S. and Nargund, V.B. (2004). *In vitro* evaluation of fungicides against *Alternaria helianthi*, causing leaf blight of sunflower. *Ind. J. Pl. Pathol.* **22(1&2) 79-82.**

Anonymous (2008). Area, Production and Productivity of Onion in India. *Indian Agri. Stat. at a Glance.*

Barnwal, M.K.; Prasad, S.M. and Mait, D. (2003). Efficacy of fungicides and bioagents against *Stemphylium* blight of onion. *Indian Phytopath.* **56 (3)**: 291-292.

Gupta, R.P. and Srivastava, P.K. (1988). Control of *Stemphylium* blight of onion bulb crop. *Indian Phytopath.* **41 (3)**: 495-496.

Kharakwal, V.P., Patni, C.S. and Adhikari, R.S. (2006). Effect of some fungicides on spore germination and growth of *Alternaria solani in vitro*. *J. Pl. Dis. Sci.*, **1 (1)**: 128-130.

Khosla, K., Thakur, B.S. and Bharadwaj, S.S. (2007). Chemical management of *Stemphylium* blight of garlic. *Pl. Dis. Res.* **22 (1)**: 47-51.

Kumari, L., Shekhawat, K.S. and Rai, P.K. (2006) Efficacy of fungicides and plant extracts against *Alternaria* blight of Periwinkle (*Catharanthus roseus*). *J. Mycol. Pl. Pathol.* **36 (2)**: 134-137.

Nene, Y.L. and Thapliyal, P.N. (1993). Evaluation of fungicides. In: *Fungicides in Plant Disease Control*, 3rd edn. Oxfrod, IBH Publishing Co., New Delhi, pp. 331.

Shahanaz, E., Razdan, V.K. and Raina, P.K. (2007). Survival, dispersal and management of foliar blight pathogen of onion. *J. Mycol. Pl. Pathol.*, **37**: 210-214.

Vincent, J.M. (1927). Distortion of fungal hyphae in the presence of certain inhibitors. *Nature*, 159-180.

Chapter 8

Screening of Antagonistic Activity of Microorganisms Against *Colletotrichum gloeosporioides*

***D.U. Gawai*[1]*, *J.U. Shinde*[1], *S.D. Jadhav*[1], *V.S. Bhadrashette*[1], *S.G. Sawant*[1] *and E.M. Khillare*[1]**

ABSTRACT

The antagonistic activities of five biocontrol agents like *Trichoderma harzianum, T. viride, Gliocladium roseum, Bacillus subtilis,* and *Pseudomonas fluorescens,* were tested *in vitro* against *Colletotrichum gloeosporioides,* the causal agents of anthracnose disease in fruit crops. The microbial antagonists inhibited mycelial growth in the dual culture assay and conidial germination *Bacillus subtilis* was evaluated under commercial packing house conditions for the control of postharvest fruit diseases of naturally infected fruit. *B subtilis,* applied in commercial Tag wax at various concentrations as well as in a water dip, significantly reduced anthracnose. The two *Trichoderma* species exhibited the strongest antagonism against *C. gloeosporioides.* Microscopic examination showed that the most common mode of action was antibiosis. The results of this study identify *T. harzianum, G. roseum, P. fluorescens,* as promising biological control agents for further testing against anthracnose disease in fruits.

Keywords: *Anthracnose, Antagonists, C. gloeosporioides, Fruit.*

---

1 Botany Research Lab. And Plant disease Clinic, N.E.S. Science College, Nanded, M.S., India

* Corresponding Author E-mail: dilip.gawai777@gmail.com

Introduction

Use of naturally occurring nonpathogenic microorganisms that are able to reduce the activity of plant pathogens and thereby suppress diseases. Antagonistic microorganisms can compete with pathogens for nutrients, inhibit pathogen multiplication by secreting antibiotics or toxins, or reduce pathogen population through hyperparasitism. Moreover, some of these microorganisms can induce generalized resistance in plants, which enables the plant hosts to better defend themselves against pathogens (Mukerji *et al.*, 2000). As a general method, biocontrol using antagonistic bacteria has been successfully demonstrated in a number of plant species. For example, bacterial isolates from the surface of banana fruits exhibited an antagonistic ability to suppress anthracnose disease caused by *C. musae* (Postmaster, 1997). Other antagonistic bacteria such as *Pseudomonas fluorescens*, *B. subtilis* and *B.megatherium* were shown to control agents of the major cotton diseases, *Xanthomonas campestris pv. malvacearum*, *Rhizoctonia solani*, *Fusarium vasinfectum* and *Verticillium dahlia* (Safiyazov *et al.*, 1995). Epiphytic bacteria isolated from apples, pears, and the surface of apple leaves showed antagonistic activity against *Penicillium expansum* (blue-mold), while *Botrytis cinerea* (gray-mold) was an effective biocontrol agent on apple fruits (Vinas *et al.*, 1997). Finally, several *Trichoderma* species can control *Colletotrichum* sp. in *Zizyphus jujuba* and anthracnose in mangos was controlled by an isolate of yeasts (Sangchote, and Saoha, 1997) but currently little is known about methods of biocontrol for anthracnose caused by *C. musae* in curcuma. Since the Bacillus genus includes some species which are known to be endophytically active (Mahaffee and Kloepper, 1997) it is likely that their endophytic ability could play a key role in the biocontrol of pathogens such as *C, musae*. Although antibiotic production by some bacteria including *Bacillus* spp. has been shown to play a major role in disease suppression(Cook *et al.*, 1995) only a few antibiotics have been isolated and identified for their role in biological control (Asaka and Shoda, 1996). Specifically two species, *B. subtilis* and *B. amyloliquefaciens* were reported effective for the control of plant pathogens (Yoshida *et al.*, 2002) due to the presence of iturin a cyclic lipo-polypeptide reported to have antibacterial activity against a large variety of yeasts and fungi (Besson *et al.*, 1979). due to the possible role of bacterial production of antibiotics in disease suppression.

The objective of this research was to investigate the biocontrol potential of antagonistic bacteria and fungi against *Colletotrichum gloeosporioides* to reduce the dependence on toxic fungicides currently required to control the diseases caused by this fungus. Investigation of alternative disease control measures is therefore urgently required. One such alternative is biological control, which has been applied successfully on several fruit and some vegetable crops (Wilson and Wisniewski, 1989).

Materials and Methods

Antagonistic Activity *in vitro*

The assay for antagonism was performed on PDA medium by a dual culture method as described by (Skidmore, and Dickinson, 1976). Briefly, the bacteria and pathogenic fungi were inoculated dually on PDA medium in petri dishes 2- 2.5 cm apart. The inhibition of actively growing fungus by the bacteria on PDA plates was quantified as the instance of radial growth in centimeters. The cultures were incubated at room temperature, and growth of the fungus towards and away from the bacterium was allowed for 7 days after incubation for each of three replicates. Variation between replicates was generally found to be less than 2mm. The percentage inhibition of the growth of the fungi was calculated using the following formula:

$$\text{Percentage inhibition of fungal growth} = (R_1 \text{--} R_2)/R_1 \times 100$$

where, R_1 was the furthest radial distance growth of the fungus in the direction of the antagonist (a control value)and R_2 was the distance on a line between the inoculation positions of the fungus and bacteria (Skidmore, and Dickinson, 1976). The antagonistic bacteria and fungi were selected for further suppression of conidial germination on *Collectotricum gloeosporioides*..

Antagonistic Inhibition on Conidial Germination of *C. gloeosporioides*

A mixture of 20 µl of *C. gloeosporioides* conidial suspension (1×10^6 cells/ml). The mixture was pipetted onto clean glass slides as a 0.5 cm diameter membrane. The slides were then incubated in moist chambers for six and a half hours using three replicate slides for each fungal isolate. At the end of the incubation period, a drop of lactophenol was added to each slide to arrest germination within the drop. Germination in this context was defined as a germ tube that had developed longer than half of the cell length. The percent germination in each fungal isolate was determined by counting 100 conidia from each isolate under the microscope and determining the proportion that had germinated.

Results and Discussion

Three fungi and two bacteria were screened for an antagonistic effect against *C. gloeosporioides* using the dual culture method described previously. The growth inhibition of *C. gloeosporioides* by all of the fungi and bacterium was determined after a week of incubation and recorded as a decay diameter (Table 8.1). All the fungi and bacteria reduced the growth diameter of the fungus by more than 40 per cent.. Two *Trichoderma* species like *T. viride* (72.60), *T. harzianum* (67.35) and *P. fluorescens* (76.45) exhibited the highest percentage of growth inhibition.. Percentage of spore germination of in culture filtrate of *C. gloeosporioides* was found more in B. Subtilis in all respective days (Table 8.2). Results obtained from the *in-vitro* study show that *P. fluorescens* could be used effectively against *C. gloeosporioides*. When the pathogen was paired with *P. fluorescens* a remarkable and almost total control of the pathogen was achieved.. It is conjectured that, its activity may be due to metabolite secreted by the bacterium into the medium.This metabolite could have brought about the change in colour of the whitish fluffy mycelia of *C. gloeosporioides* to purple. Similarly, the work carried out by different workers by using antagonists. *T. harzianum* has been efficient in control of several pathogens (Adekunle *et al.*, 2001). The potential values of *Trichoderma* spp as bioagents were reported by Howell, (2003) for the protection of several seedlings, potted outdoor and field diseases of crops. Mohammed and Amusa (2003) observed that *T. harzianum* grew over all the pathogens tested in their study. Mechanisms of antagonism suggested by other researchers were mycoparasitism

(Elad *et al.*, 1983) and rhizosphere competence (Howell, 2003), which are the factors responsible for hyperparasitism. Bourah and Kumar (2003) reported that the secondary metabolites produced by a strain of *P. fluorescens* produce antibiotics phenazine (PHE) 2, 4-diacetyl phloroglucinol (PHL) and siderophone phyoverdin (PYO) in king's B and succinic acid media respectively. Pukall *et al.* (2005) identified four different types of *Bacillus* spp. namely *B. pumils*, *B. fusiforms*, *B. subtilis* and *B. mojavensis* with toxin producing strains outside *B. cereus*.

Table 8.1: Inhibitory Rate in Percentage (per cent) with the Respect to Incubation Days

Sl.No.	*Antagonist*	*Incubation Days*			
		2 Days	*4 Days*	*6 Days*	*8 Days*
1.	*T. viride*	18.90	32.68	63.45	72.60
2.	*T. harzianum*	14.00	28.74	54.36	67.35
3.	*G. roseum*	13.80	17.40	30.12	41.26
4.	*B. subtilis*	19.90	24.40	36.00	46.75
5.	*P. fluorescence*	25.90	38.35	66.25	76.45

Table 8.2: Percentage of Spore Germination of in Culture Filtrate of *C. gloeosporioides*

Sl.No.	*Antagonists*	*% of Spore Germination in Culture Filtrate at Different Days*			
		2 Days	*4 Days*	*6 Days*	*8 Days*
1.	*T. viride*	06	10	15	18
2.	*T. harzianum*	09	15	20	25
3.	*G. roseum*	15	20	24	30
4.	*B. subtilis*	20	35	40	50
5.	*P. fluorescence*	07	12	18	20

In vitro growth of *C. gloeosporioides* was found to be markedly suppressed by bacterial species: *B. subtilis*. Out of the three fungal species screened for *in vitro* growth suppression of the fungus, these three species exhibited the highest level of fungal growth inhibition (over 75 per cent) and at concentrations over $2x10^8$ CFU/ml provided an absolute inhibitory effect on the germination of this fungus. Therefore assuming that sufficient levels of the bacteria could be induced to colonize the plants, the *Pseudomonas* species described here could have the potential to effectively treat and stop the spread of fungal infections by *C. gloeosporioides*. A dual culture test showing growth suppression of *C. gloeosporioides*. This could be due in part to an antibacterial compound in *P. fluorescence* that inhibited the growth of *C. gloeosporioides*.These findings are supported by the work of Bourah and Kumar(2002).

Bacillus subtilis was the least effective organism used as antagonist in the control of *C. gloeosporioides* because of its inability to significantly inhibit the growth of the

later. *B. subtilis* is generally known to inhibit growth of other microbes by antibiosis. The restriction observed on the growth of *C. gloeosporioides* in this *in vitro* study might be due to the rapid colonization of the growth substrate by *B. subtilis*, which had a faster growth rate than the *P. fluorescence* into the growth medium. It stopped the hyphal growth of *C. gloeosporioides* even though the effect of the antibiotics was not noticed in the *B. subtilis*. It is also clear from the results that very less percentage of spore germination recorded in culture filtrate of *T.viride, T.harzianum* and *P. fluorescence*. It may be due to inhibitory substances produced by these antagonists. This investigation clearly showed that postharvest application of antagonistic *B subtilis* controlled anthracnose and establishes biocontrol as a viable alternative to the use of chemicals for the control of anthracnose disease caused by C. *gloeosporioides*

Acknowledgements

Authors are thankful to Principal, N.E.S. Science College, Nanded for providing facilities

References

Adekunle, A.T., Cardwell, K.F., Florini, D.A. and Ikotun, T. (2001). Seed treatment with *Trichoderma* species for control of Damping-off of Cowpea caused by *Macrophomina phaseolina. Biocontr. Sci. Technol.*, **11**: 449-457.

Asaka, O. and Shoda, M. (1996). Biocontrol of *Rhizoctonia solani* damping-off of tomato with *Bacillus subtilis* RB14. *Appl. Environ. Microbiol.*, **62**: 4081-4085.

Besson, F., F. Peypoux, G. Michel and Delcambe, L. (1979). Antifungal activity upon derivatives; inhibition of this antifungal activity by lipid antagonists. *J. of Antibiot.*, **32**: 828-833.

Bourah, H.P.D.and Kumar, B.S.D. (2002). Biological activities of secondary metabolite produced by a strain of Pseudomonas fluorescence. *Foliamicrobiologica, Phali Czech Republic* **47**(4): 359-365.

Cook, R.J., L.S. Thomashow, B.M. Weller, D.Fujimoto, M. Mazzola, G. and Bangera, D.S..Kim, (1995). Molecular mechanisms of defense by rhizobacteria against root disease.–*Proc.of the Nat. Acad. of Sci. of the USA* **92**: 4197-4201.

Elad, Y.I., Chet, P., Boyle and Henis Y. (1983). Parasitism of *Trichoderma* spp on *Rhizoctonia solani* and *Sclerotinum rolfsii*-scanning electron microscopy and fluorescence microscopy. *Phytopathology*. **73**: 85-88.

Howell, C.R. (2003). Mechanisms employed by *Trichoderma* species in the biological Control of Plant diseases: the history and evolution of current concepts. *Plant Dis*,. **87(1)**: 4-10.

Mahaffee, W.F., and J.W. Kloepper (1997). Temporal changes in the bacterial communities of soil, rhizosphere and endorhiza associated with field-grown cucumber (*Cucumis sativus* L.). *Microb. Ecol.*, **34**: 210-223.

Mohammed, S, and Amusa, N.A. (2003). *In vitro* inhibition of growth of some seedling blight inducing pathogens by compost-inhabiting microbes. *African journal of Biotechnology* cited online www.Academicjournals.org/AIB manuscripts/ manuscripts 2003 (15/12/2004).

Mukerji, K.G., B.P.Chamola and R.K.Upadhyay (2000). Biotechnological approaches in biocontrol of plant pathogens. *Phytochem.*, **54**: 445-446.

Postmaster, A., J. Kuo, K. Sivasithamparam, and D.W. Turner (1997). Interaction between *Colletotrichum musae* and antagonistic microorganisms on the surface of banana leaf discs. *Scient. Horti.* **1**: 113-125.

Pukall C.R., Schumann. P., Hormazabal. V., Granum. P. (2005). Toxin producing ability among *Bacillus* spp. Outside *Bacillus cereus* group. *Appl. Environ. Microbiol.* **71(3)**: 1178-1183.

Safiyazov, J.S.,R.N.Mannanov and R.K.Sattarova (1995).The use of bacterial antagonists for the control of cotton diseases. *Field Crop Research* **43**: 51-54.

Sangchote, S. and M. Saoha (1997). Control of Postharvest Disease of Mango Using Yeasts.–Aciar Proc. 81: 108-112. *Am. J. Agril. and Biol. Sci.*, **2 (2)**: *54-68.*

Skidmore, A.M. and C.H. Dickinson, 1976. Colony interaction and hyphal interference between *Septoria nodorum* and phylloplane fungi. *Trans. Br. Mycol. Soc.* **66**: 57-64.25.

Vinas, I., J. Usall, N. Teixido and V. Sanchis (1997). Biological control of major postharvest pathogens on apple with *Candida sake. J. of Food Microbio.* **40**: 9-16.

Wilson, C.L. and Wisniewki, M.E. (1989). Biological control of post harvest diseases of fruits and vegetables: An emerging technology. *Ann. Rev. of Phytopath.*, **27**: 425-441.

Yoshida, S., Shirata, A. and Hiradate, S. (2002). Ecological characteristic and biological control of mulberry anthracnose. *JARQ*, **36**(2): 89-95.

Chapter 9

Biopriming of Seeds for Improving Germination Behaviour and Biomass Production of Oilseed Crops

J.N. Rajkonda[1], S.L. Korekar[1], V.S. Sawant[2] and U.N. Bhale[2]

ABSTRACT

Crop seeds of groundnut (*Arachis hypogaea*), soybean (*Glycine max*) and mustard (*Brassica campenstris*) were primed with five species of *Trichoderma* using spore suspension to asses their effect in improving germination of seeds, vigour index and seedling biomass of crops. Significant enhancement in the seed germination (per cent), vigour index and seedling biomass were noted. The spore suspension of each *Trichoderma* species enhanced the germination (per cent), vigour index and biomass production over control. *Trichoderma pseudokoningii* showed higher germination (95 per cent) and vigour index (1520) on groundnut followed by *T. virens*. However, *T. koningii* showed significant effect on biomass production of groundnut (0.70 gm.). On the soybean, germination (90 per cent), And vigour

1 Department of Botany, Yeshwantrao Chavan Mahavidyalaya, Tuljapur Dist. Osmanabad – 413 601, M.S., India, E-mail: jnrajkonda@gmail.com

2. Research Laboratory, Department of Botany, Arts, Science and Commerce College, Naldurg, Dist. Osmanabad – 413 602, M.S., India, E-mail: unbhale2007@rediffmail.com

index (2610) was accelerated by *T. virens* followed by *T. pseudokoningii*, *T. viride* and *T.harzianum*. Biomass production was recorded high when *T. viride* was treated. Germination (85 per cent) of mustard was high in the treatment of *T. koningii*. Vigour index (872.2) and biomass production (7.07gm) of mustard was higher under the treatment of *T. harzianum*.

Keywords: *Biopriming, Bioagents, Oilseed crops, Trichodrerma spp.*

Introduction

Seed treatment with bio control agents for protection of seeds and control of seed borne diseases offers farmers an alternative the use of chemical fungicides. The seed treatment with bioagent can be highly effective. These are more efficient than chemical seed protectants. Some biological control agents used as seed treatment were capable of colonizing the rhizosphere and providing benefits to plants beyond the emergence stage of the seedlings (Callan *et al.*, 1997).

The efficacy of biological seed treatment can also be affected by soil pH and iron concentration (Weller, 1998) and moisture, temperature and inoculums density of the pathogen (Mathre *et al.*, 1994). Efficacy can also be influenced by certain characteristics of the bio control agent and of the seed treatment itself. These include the inoculum density of the biocontrol agent on the seed adjunct treatment such as priming formulations and pathogen specificity that enhance the activity and survival of the biocontrol agent in the formulated product, crop and pathogen specificity of the biocontrol agent and compatibility with their microbial inoculants or chemical fungicides. Seed treatment encompasses a variety of physiological treatments that enhance seed germination and vigour through the addition of moisture (Heydecker and Coolbear 1977; Bennet and Waters 1987; Taylor *et al.*, 1988; Khan, 1992). The addition of biocontrol agents during biopriming allows for colonization of the seed prior to planting and adds a new dimension to seed priming treatment. Therefore, a present study was made to test the efficacy of *Trichoderma* species in improving germination behavior of some seeds of oil crops.

Materials and Methods

The biopriming of seeds of oilseed crops like groundnut (*Arachis hypogeal*), soybean (*Glycine max*) and mustard (*Brassica campenstris*) was done with five species of *Trichoderma viz- T. viride, T. harzianum, T. koningii, T. pseudokoningii* and *T. virens*. All the species of *Trichoderma* were obtained from rhizosphere soil of different crop plants from the Marathwada region of Maharashtra. Twenty healthy seeds of each crop plant were selected and treated with the *Trichoderma* species separately. The treated seeds were sown in separate pots containing 2 kg sterilized black soil. The seeds without treatment were also sown in another pot and considered as control. Water was poured with regular intervals to each pot. Observations made in terms of per cent germination, shoot length, root length, fresh weight, dry weight and number of leaves. The vigour index of seedlings in each crop was also calculated (Abdul Baki and Anderson, 1973) Vigour index = [root (cm) +shoot (cm)] x germination (per cent).

Results and Discussion

In present investigation there was increase in per cent germination, biomass production and vigour index of test crops by the treatment of *Trichoderma* species. *Trichoderma pseudokoningii* was effective in inducing germination (95 per cent) as well as increasing the number of leaves and vigour index of groundnut. However, the root length was more by *T. harzianum* while shoot length, fresh weight and dry weight were increased by *T. koningii* (Table 9.1).

Table 9.1: Effect of Seed Priming with *Trichoderma* Species on Germination Biomass Production and Vigour Index of Groundnut (*Arachis hypogea*)

Sl.No.	Source	Germination (%)	No. of Leaves	Root Length cm	Shoot Length cm	Fresh Weight gm	Dry Weight gm	Vigour Index
1.	Control	45 per cent	6.67	6.52	4.81	8.37	0.32	509.8
2.	*T. viride*	65 per cent	4.77	6.07	3.70	7.62	0.35	635.0
3.	*T. harzianum*	50 per cent	7.40	11.10	6.16	8.71	0.37	863.0
4.	*T. koningii*	55 per cent	7.33	10.81	7.00	14.73	0.70	979.5
5.	*T. pseudokoningii*	95 per cent	8.22	9.74	6.26	9.30	0.38	1520.0
6.	*T. virens*	85 per cent	7.53	11.00	5.91	8.02	0.32	1437.8

Trichoderma virens was more effective in germination (90 per cent) of soybean followed by *T. viride, T. harzianum* and *T. pseudokoningii.* The number of leaves was more when *T.viride* was treated. Shoot length was more under the treatment of *T. pseudokoningii* while root length was increased by *T.virens.* Increased fresh weight was recorded by *T.virens* while dry weight was more under the treatment of *T. viride.* Vigour index was recorded maximum by the treatment of *T. virens* (Table 9.2).

Table 9.2: Effect of Seed Priming with *Trichoderma* Species on Germination Biomass Production and Vigour Index of Soybean (*Glycine max*)

Sl.No.	Source	Germination (%)	No. of Leaves	Root Length cm	Shoot Length cm	Fresh Weight gm	Dry Weight gm	Vigour Index
1.	Control	55 per cent	5.81	11.46	11.67	7.82	0.30	1272.1
2.	*T. viride*	85 per cent	6.12	10.84	13.34	7.61	0.37	2055.3
3.	*T. harzianum*	85 per cent	6.00	10.37	13.07	6.16	0.18	1992.4
4.	*T. koningii*	75 per cent	7.33	10.03	13.65	7.71	0.28	1776.0
5.	*T. pseudokoningii*	85 per cent	5.83	12.39	16.65	7.87	0.33	2468.4
6.	*T. virens*	90 per cent	5.81	14.09	14.91	7.96	0.34	2610.0

The effect of *T. koningii* was high on the germination (85 per cent) of mustard. *Trichoderma viride* and *T. harzianum* were equally effective on germination (75 per cent). The shoot length was noted maximum by the treatment of *T. pseudokoningii.* However, the *T. harzianum* was significant with respect to root length, fresh weight, dry weight, number of leaves and vigour index (Table 9.3).

Table 9.3: Effect of Seed Priming with *Trichoderma* Species on Germination Biomass Production and Vigour Index of Mustard (*Brassica campenstris*)

Sl.No.	Source	Germination (%)	No. of Leaves	Root Length cm	Shoot Length cm	Fresh Weight gm	Dry Weight gm	Vigour Index
1.	Control	45 per cent	5.33	4.93	3.28	0.337	0.043	369.9
2.	*T. viride*	75 per cent	6.00	5.96	3.55	0.382	0.047	714.0
3.	*T. harzianum*	75 per cent	7.07	11.63	7.59	1.436	0.205	872.2
4.	*T. koningii*	85 per cent	6.00	9.06	5.86	0.316	0.035	770.1
5.	*T. pseudokoningii*	60 per cent	6.00	11.33	6.14	0.459	0.054	679.8
6.	*T. virens*	55 per cent	6.12	10.67	6.91	0.500	0.054	586.8

Our present results were in accordance with the earlier findings of several researches (Khan 1992; Sung and Chang, 1993) where they concluded enhanced seed germination and seedling vigour during the seed priming. Pre-colonization provides the biocontrol agent with a competitive advantage over attacking pathogens and often provides superior seed protection when compared to seed coating (Harman and Taylor 1988; Harman *et al.*, 1989; Harman 1991). Prasad *et al.* (2002) had reported that biopriming with *Trichoderma* resulted into increased germination, root and shoot length of red gram. Increased biomass production was observed with tobacco seedlings by treatment with *T. harzianum* conidia (Chacon *et al.*, 2007). *Trichoderma* species can compete with other microorganisms for key exudates from seeds that stimulate the germination of propagules of plant pathogenic fungi in soil (Howell 2003) and generally compete with soil microorganisms for nutrients and/or space (Elad, 1996). Recently Bhagat and Pan (2007) were recorded increases in seed germination, vigour index and seedling biomass of treated seeds of crop plants.

Conclusion

In the present study, there was a increase in germination (per cent), root length, shoot length, biomass production and seedling vigour index by the treatment of *Trichoderma* species. The seedlings were healthy and disease free. Among the *Trichoderma* species, *T. pseudokoningii* was more effective on the groundnut seeds. The effect of *T. virens* was maximum on soybean seeds. However, *T. harzianum* was more effective on mustard seeds.

Acknowledgement

Authors are thankful to UGC, WRO (PUNE) for financial assistance.

References

Abdul Baki, A.A. and Anderson, S.P. (1973). Vigour determination in soyabean seeds by multiple criteria. *Crop Sci.*, **13**: 630-633.

Bennet, M.A. and Waters, I. (1987). Seed hydration treatment for improved sweet corn germination and stand establishment. *J. Americ. Soc. Hort. Sci.*, **112**: 45–49.

Bhagat, S. and Pan, S. (2007). Mass multiplication of *Trichoderma harzianum* on agricultural byproducts and their evaluation against seedling blight of mungbean and collar rot of groundnut. *Indian J. Agric. Sci.*, **77**: 583–588.

Callan, N.W., Mathre, D.E., Miller, J.B. and Vavrina, C.S. (1997). Biological seed treatments: factors involved in efficacy. *Hort. Sci.*, **32**: 179–183.

Chacon, M.R., Rodriguez-Galan, O., Benitez, T., Sousa, S.,Rey, M., Llobell, A.and Delgado-Jarana, J. (2007). Microscopic and transcriptome analysis of early colonization of tomato roots by *Trichoderma harzianum*. *Int. Microbiol.*, **10**: 19–27.

Elad, Y. (1996). Mechanisms involved in the biological control of *Botrytis cinerea* incited diseases. *Eur. J. Pl. Pathol.*, **10**: 2719-72.

Harman, G.E. and Taylor, A.G. (1988). Improved seedling performance by integration of biological control agents at favourable pH levels with solid matrix priming. *Phytopathology.*, **78**: 520-525.

Harman, G.E., Taylor, A.G. and Stasz, T.E. (1989). Combining effective strains of *Trichoderma harzianum* and solid matrix priming to improve biological seed treatments. *Pl. Dis.*, **73**: 631-637.

Harman, G.E. (1991). Seed treatments for biological control of plant disease. *Crop Prot.*, **10**: 166-171.

Heydecker, W. and Coolbear, P. (1977). Seed treatments for improved performance-survey and attempted prognosis. *Seed Sci. Tech.*, 5: 353–425.

Howell, C.R. (2003). Mechanisms employed by *Trichoderma* species in the biological control of plant diseases: The history and evolution of current concepts. *Pl. Dis.*, **87**: 4-10.

Khan, A.A. (1992). Preplant physiological seed conditioning. *Hort. Rev.*, **13**: 131-181.

Mathre, D.E., Callan, N.W., Johnson, R.H., Miller, J.B. and Schwend, A. (1994). Factors influencing the control of *Pythium ultimum*-induced seed decay by seed treatment with *P. aureofaciens*. *Crop Prot.*,**13**: 301–307.

Prasad, R.D., Rangeshwaram, R., Hegde, S.V. and Anuroop, C.P. (2002). Effect of soil and seed application of *Trichoderma harzianum* on pigeonpea wilt caused by *Fusarium udum* under field conditions. *Crop Prot.*, **21**: 293–297.

Sung, F.J.M. and Chang, Y.H. (1993). Biochemical activities associated priming of sweet corn seeds to improve vigour. *Seed Sci. Tech.*, 21: 97 105.

Taylor, A.G., Klein, D.E. and Whitlow, T.H. (1988). SMP: Solid matrix priming of seeds. *Scient Hort.*, **37**: 1–11.

Weller, D.M. (1988). Biological control of airborne plant pathogens in the rhizosphere with bacteria. *Ann. Rev. Phytopathol.*, **26**: 379–407.

Chapter 10

Mycorrhiza: Biofertilizer for *Solanum melongena* L.

V.S. Sawant[1], P.P. Sarwade[1], V.S. Chatage[1], J.N. Rajkonda[2] and U.N. Bhale[1]*

ABSTRACT

The significance of Arbuscular Mycorrhizal (AM) fungi on plant growth has been recognized. AM fungi status was assessed in pot culture using sterile soil. AM fungi were isolated from soil of Osmanabad district of Marathwada, Maharashtra. Biomass parameters of *Solanum melongena* L. was examined after 90 days of sowing.. No mannuring was done during this period. Indigenous consortium of AM fungi was applied alone and also with *Trichoderma viride* and *T. harzianum*. AMF inoculation significantly increased root dry weight (1.10 g) and shoot dry weight (2.37 g) over control. When indigenous consortium of AMF was inoculated with *T. viride* root dry weight (1.13g) and shoot dry weight (2.53g) was increased. In case of *T. harzianum,* root dry weight (1.26 g) and shoot dry weight (2.34 g) also shows increased growth over control.

Keywords: AM fungi, Growth parameters, Solanum melongena L.

---

1 Research Laboratory, Department of Botany, Arts, Science and Commerce College, Naldurg, Tal. Tuljapur, Dist. Osmanabad – 4136 02, M.S. India

2 Department of Botany, Yeshwantrao Chavan College, Tuljapur, Dist. Osmanabad – 413 601, M.S., India

* Corresponding Author E-mail: vjysawant@gmail.com

Introduction

In India Brinjal (*Solanum melongena* L.) fruits are widely used as a vegetable. Association of Arbuscular Mycorrhizal (AM) fungi with brinjal have been studied by different workers (Ramchandra and Rai, 1987 and Edathil *et al.*, 1999). AM fungi increase performance in low nutrient soil (Draft and Nicolson, 1966 and Abbot and Robson, 1982). This work deals with biomass production using mycorrhiza and without organic or inorganic mannuring. Biomass and biochemical constituents of plants get increased when inoculated with AM fungi in fruit crops like mango, papaya, banana, jamun and citrus (Patil and Patil, 2007). Due to beneficial effect, arbuscular mycorhizae are receiving considerable attention in agriculture and forestry (Peterson *et al.*, 1984). Bagyaraj and Sreeramulu (1982) found increase in height and flowering in chilli plant. Dry matter in pepper plants heavily increased when supplemented with *Glomus intraradices* (Semra Demir, 2004). Martinez *et al.* (2004) has also studied beneficial effect of micorrhiza and *T. pseudokoningii.*

Materials and Methods

The commercial Brinjal (*Solanum melongena* L) variety was used for study. The experiment was conducted at Department of Botany, Arts, Science and Commerce College, Naldurg. Soil chemical analysis prior to experiment was determined. Total phosphorus (P) was calorimetrically determined and total nitrogen (N) was determined by micro Kjeldhal digestion and measured according to Jackson (1971). Potassium (K) was examined by a digital flame photometer. Twelve pots of 18 cm diameter were filled with incubated soil. Indigenous AM flora was collected from brinjal rhizosphere soil by wet sieving and decanting method (Gerdemann and Nicolson, 1963). Five holes were made in pot soil up to three cm depth and AM propagules were inoculated. Surface sterilized brinjal seeds were placed on it and holes were filled with sterile soil. Three replicates were maintained of each treatment. After 45 days four plants per pot were removed and used for root colonization. One plant per pot was maintained for 90 days. After 90 days plants were carefully uprooted. Length of root and shoot, fresh and dry weight were measured. The indigenous AM flora consisted of *Glomus aggregatum* Schenk and Smith emend Koske, *Acaulospora scorbiculata* Trappe, *G. fasciclatum* (Thaxter) Gerd. And trappe emend Walker and Koske, *A. tuberculata* Janes and trappe and *G. mosseae* Nicol. and Gerd.) Gerd. and trappe. Statastical analysis of the experiment were performed by Mungikar (1997).Microbial inoculation effect (MIE) was determined according to Bagyaraj (1992).

$$MIE = \frac{\text{Dry weight of inoculated plant * – Dry weight of noninoculated plant}}{\text{Dry weight of inoculated plant}} \times 100$$

* includes all microbial inoculations including AM fungi.

Result and Discussion

All plants except control were colonized by mycorhiza at 45 days. Colonization showed vescicles, hyphae and arbuscles. The arbuscles are the structures involved in the transfer of nutrients between fungal and host plants (Cox and Tinker, 1976).

The shoot dry weight of mycorrhizal plant (2.73 g) was higher than non mycorrhizal plant (1.77 g). There is increase in dry shoot biomass when indigenous mycorrhiza was implemented along with *T. viride* (2.53 g) and *T. harzianum* (2.34 g) (Table 10.2). Sreeramulu and Bagyaraj (1986) observed increase in shoot dry weight after addition of mycorhiza in chilli plant. When mycorrhiza was supplemented with *T. viride* it showed better yield. (Sarwade, 2011). Fresh root wt. (5.83 g) and dry root wt. (1.26 g) was highest when indigenous mycorrhiza was used with *T. harzianum* as a compared to control (2.05 g and 0.53g) and mycorrhiza alone (3.14 g and 1.10 g) respectively. Enhanced growth of root system helps in more nutrient absorption resulting in increase of plant biomass (Tanwar *et al.*, 2010). The tested soil was determined chemically in this experiment, pH was 7.98, EC 0.89, organic carbon 0.39 per cent and have more nitrogen 279.10 kg/h than P and K (Table 10.1). Based on the result it is concluded that it is possible to increase biomass through inoculation with indigenous mycorrhiza and easily available *Trichoderma* isolates.

Table 10.1: The Chemical Analysis of Soil Used in the Study for Pot Experiment

pH	*EC*	*Organic Carbon per cent*	*N kg/h*	*P kg/h*	*K kg/h*
7.98	0.89	0.39	279.10	45.58	224.00

Table 10.2: Plant Height, Fresh and Dry Weight of Brinjal Plant 90 Days After Sowing in Pot Experiment

Treatment	*Root Length cm*	*Shoot Length cm*	*Root Fresh Weight g*	*Root Dry Weight g*	*Shoot Fresh Weight g*	*Shoot Dry Weight g*	*Plant Dry Weight g*	*MIE %*
Indigenous Mycorrhiza	14.6	20.9	3.14	1.10	8.50	2.37	3.47	33.72
Indigenous Mycorrhiza+ *T. viride*	15.2	18.3	5.25	1.13	11.66	2.53	3.66	37.16
Indigenous Mycorrhiza+ *T. harzianum*	19.6	16.3s	5.83	1.26	9.08	2.34	3.60	36.11
Control	13.4	15.9	2.05	0.53	6.93	1.77	2.30	–
SE ±	1.35	1.14	0.89	0.16	0.98	0.17	0.32	–
CD at 5%	4.30	3.64	2.82	0.52	3.13	0.53	1.02	–

Acknowledgements

The authors are thankful to UGC WRO Pune for the financial assistances of Minor Research Project and also grateful to the Principal, Dr. S.D. Peshwe for providing laboratory facilities.

References

Abbott, L.K. and Robson, A.D. (1982). The role of vesicular-arbuscular mycorrhizal fungi and selection of fungi for inoculation. *Australian Journal of Agriculture Research*, **33**: 389-408.

Alicia Martinez, Mariana Obertello, Alejandro Pardo, Juan A. Ocampo, Alicia Godeas (2004). Interactions between *Trichoderma pseudokoningii* strains and the arbuscular mycorrhizal fungi *Glomus mosseae* and *Gigaspora rosea*. *Mycorrhiza*, **14**: 79– 84.

Bagyaraj, D.J. (1992). Vesicular arbuscular mycorrhizae; Application in agriculture. In: *Methods in Microbiology*, (Eds.) J.R. Norris., D.J. Read and A.K. Varma, Academic Press, pp. 359-373.

Bagyaraj, D. J. and Sreeramulu, K. R. (1982). Preinoculation with VA mycorrhiza improves growth and yield of chilli transplanted in the field and saves phosphatic fertilizer. *Plant and Soil*, **69**: 375-381

Cox, G.C. and Tinker, P.B. (1976). Translocation and transfer of nutrients in vesicular arbuscular mycorrhizas. The arbuscle and phosphorus transfer: a qualitative ultrastructural study. *New Phytol.*, **73**: 901-912.

Draft, M.J.and Nicolson, T.H. (1966). Effect of endogone mycorrhiza on plant growth. *New Phytol.* **65**: 343-350.

Edathil, T. T., Manian, S. and Udayan, K. (1996). Interaction of multiple VAM fungal species on root colonization, plant growth and nutrient status of tomato seedlings (*Lycopersicon esculentum* Mill.). *Agriculture, Ecosystems and Environment* **59**: 63-68

Gerdemann, J.W. and Nicolson, T.H., 1963. Spores of mycorrhizal Endogone species extracted from soil by wet sieving and decanting. *Trans. Br. Mycol. Soc.*, 46: 235-244.

Jackson, M.L. (1971). *Soil Chemical Analysis*. Prentice Hall, New Delhi, pp. 498.

Mungikar, A.M. (1997). *An Introduction to Biometry*. Saraswati Printing Press, Aurangabad, pp. 57-63.

Peterson, R.L., Piche, V. and Plenchette, C. (1984). Mycorrhizae and their potential use in agriculture and forestry industries. *Biotech. Adv.*, **2**: 110-120.

Patil, P. B. and Patil, C.P. (2007). Mycorrhizal biotechnology for increasing growth and productivity of fruit plants. In: *The Mycorrhizae: Diversity Ecology and Application*. Daya Publishing House, pp. 57-86.

Ramchandra, G.S. and Rai, P.V. (1987). Effect of *Azatobactor chrococcum* and *Glomus fasciculatum* on growth and yield of brinjal (*Solanum melongena* L.). *Indian J. Microbiol.*, **27**: 78-80.

Sarwade, P.P. (2011). Studies on arbuscular mycorrhizal fungi in selected medicinal plant species of Marathwada. *Ph.D. Thesis*, Dr. Babasaheb Ambedkar Marathwada University, Aurangabad.

Semra Demir (2004). Influence of arbuscular mycorrhiza on some physiological growth parameters of pepper. *Turk J. Biol.*, **28**: 85-90.

Sreeramulu, K.R. and Bagyaraj, D.J. (1986). Field response of chilli VA mycorrhiza on black clayey soil. *Plant and Soil*, **93**: 299-302.

Tanwar, A. Kumar, A. Mangala, C. and Aggarwal, A. (2010). Effect of AM fungi and *Trichoderma herzianum* on growth response of *Lycopersicum esculentum*. *J. Mycol. Pl. Pathol.*, **40(2)**: 219-223.

Chapter 11

Efficacy of Indigenous AM Fungi on Growth and Nutrient Uptake of a Ground Orchid– *Habenaria marginata* Colber.

A. Channabasava[1] and H.C. Lakshman[1]*

ABSTRACT

In most of the orchids, mycorrhizae are crucial for germination of the seed and further development of the seedlings. In the present work six different native AM fungal mixed inoculum were screened to understand the growth responses and nutrient uptake of ground orchid. The plant biomass in terms of dry weight of shoot and root and P uptake in shoot was measured in all the treated and control plants. Results revealed that the improvement of overall growth of *Hebanaria marginata* Colber., was noted. The plants showed significantly increased growth and nutrient uptake, when they are inoculated with *Glomus macrocarpum* followed by the inoculation of *Scutellospora erythropa* and *Glomus bagyarajii* compared to remaining three AMF treated and control plants. The significantly increased per cent of root colonization and spore number was recorded in mycorrhizal

---

1 P.G. Department of Studies in Botany (Microbiology lab), Karnatak University, Dharwad – 580 003, Karnataka, India

* Corresponding Author E-mail: casahukar85@gmail.com

plants over the control plants. In the present investigation the effect of different AM fungi on growth and nutrient uptake by *H. marginata* Colber. has been discussed.

Keywords: *Hebanaria marginata, Glomus macrocarpum, Per cent root colonization, Spore number, Nutrient uptake, Biomass production.*

Introduction

Living together is one of the most prevalent phenomenons in the biological world, especially in the plant kingdom and in the underground environment. Underground world also harbors one of the most common symbiotic associations between plant root and fungus called Mycorrhiza (Smith and Read, 1997; Verma *et al.*, 2002). Mycorrhiza represents an important group because they have a wide distribution; may contribute significantly to microbial biomass and to soil nutrient cycling processes in plants (Harley and Smith, 1983).

Arbuscular Mmycorrhizal Fungi (AMF) benefit their host principally by increasing uptake of relatively immobile phosphate ions, due to the ability of the fungal ERM to grow beyond the phosphate depletion zone that quickly develops around the root. In return, the fungi receive carbon (C) from the host plant. Other benefits to the host that have been identified include: increased resistance to foliar-feeding insects, improved drought resistance, increased resistance to soil pathogens (Sylvia and William, 1992; Davet, 1996; Smith and Read, 1997) and increased tolerance of salinity and heavy metals. Increased uptake of macronutrients other than P, including nitrogen (N) potassium (K) and magnesium (Mg) has also been measured (Read, 1990) as well as increased uptake of some micronutrients. Mycorrhizal associations are beneficial to plants and thus crop productivity for sustainable agriculture (Gianinazzi- Pearson and Diem, 1982; Bethlenfalvay, 1992). In addition, mycorrhizae have been shown to play an important role in maintaining soil aggregate stability. Though the AM association can offer multiple benefits to the host plant it may not be obviously mutualistic at all points in time, and it is possible under some conditions that the AMF may cheat their host plant into supplying C with no apparent benefit to the plant.

Perennial plants which are members of the largest family of plants–Orchidaceae are known as Orchids. They are known for their longer lasting and bewitchingly beautiful flowers which fetch a very high price in the international market. *Habenaria* species have small to large underground root tubers. The flowers are mostly green, white and yellow. Petals are either unlobed or deeply two-lobed. Lips are usually three-lobed and have a long spur that is sometimes swollen at the end. Research works on terrestrial orchids with inoculation of AM fungi are very scanty. Therefore present investigation was undertaken to screen the efficient AM fungus to understand the growth responses in terrestrial orchid *Habenaria marginata.*

Material and Methods

Twenty days old plantlets of terrestrial orchid *Habenaria marginata* Colber. were collected from the Botanical garden, Karnatak University, Dharwad (INDIA) at early stage during the month of September and October-2010. They were transplanted in pots, measuring about 12x15 cm diameter containing 4 kg of garden soil with pH 6.7. The control plants are not inoculated with any AMF inoculum. The six different AM fungal species were recovered by adapting wet sieving and decanting method (Gerdeman and Nicolson, 1963) from the rhizospheric soils. The isolated spores of AM fungi were mass multiplied by using *Sorghum vulgarae* L. as a host plant in separate earthen pots containing sterilized soil and sand mixture (3:1). 10 g of inoculum containing dry soil, hyphae, spores and root bits was mixed in the top of 8 cm of the each experimental pot. The experimental pots were arranged in randomized complete block design with triplicates for each treatment. The treatments are as Soil without AM inoculums, soil with *Glomus macrocarpum* (GM), soil with *Glomus bagyarajii* (GB), soil with *Scutellospora erythropa* (SE), soil with *Sclerocystis dussii* (SD), soil with *Gigaspora marginata* (MG) and soil with *Acaulospora laevis* (AL).

The plants were harvested to determine the per cent of root colonization (Philips and Hayman, 1970) and spore number in the rhizospheric soils. The plant height and biomass were recorded and the percent of 'P' content in shoot also determined (Jackson, 1973).

Table 11.1: The Effect of Different AM Fungi on Growth and Nutrient Uptake of *Habenaria marginata* Colber

Treatments	*PH*	*DWS*	*DWR*	*P*	*PMC*	*SN*
GM	84.93±0.11a	14.65±0.16a	4.05±0.04a	2.53±0.11a	94.33±2.33a	192±1.73a
GB	65.81±0.22b	11.67±0.15b	2.78±0.08b	1.99±0.02b	86.00±1.00b	169.6±1.45b
SE	70.09±0.15c	10.45±0.18c	2.43±0.08c	1.86±0.03b	81.33±1.45b	147.6±1.76c
SD	59.04±0.09d	9.33±0.08d	1.97±0.03d	1.65±0.04c	73.66±1.85c	124.6±2.60d
MG	54.68±0.20e	8.03±0.05e	1.69±0.02e	1.49±0.02d	73.00±0.57c	113.6±2.18e
AL	49.9±0.12f	6.62±0.10f	1.63±0.07e	1.32±0.02e	71.66±2.18c	100.6±1.20f
CN	33.36±0.25g	4.83±0.10g	1.24±0.01f	1.06±0.02f	0.00±0.00d	0.00±0.00g

PH: Plant Height; DWS: Dry Weight of Shoot; DWR: Dry Weight of Root; P: Phosphorus; PMC: Per cent of Mycorrhizal Colonization; SN: Spore Number

GM: *Glomus macrocarpum;* GB: *Glomus bagyarajii;* SE: *Scutellospora erythropa;* SD: *Sclerocystis dussii;* MG: *Gigaspora marginata;* AL: *Acaulospora laevis;* CN: Control.

Each value represents the mean of three determinations. Mean values followed by the same letter within a column do not differ significantly at P = 0.05 according to DMRT.

Results

In the present investigation six different AM fungal species were used as inocula to understand the growth responses of the *H. marginata*. Plants showed significantly increased growth and nutrient uptake, when they are inoculated with arbuscular mycorrhizae over the control plant. There was highest increase in plant height (Figure 11.2), dry weight of both shoot and root were recorded in the plants treated with *G. macrocarpum* and it was followed by inoculation of *S. erythropa* and *G. bagyarajii* compared to remaining three AMF inoculated and control plants (Table 11.1). Plant

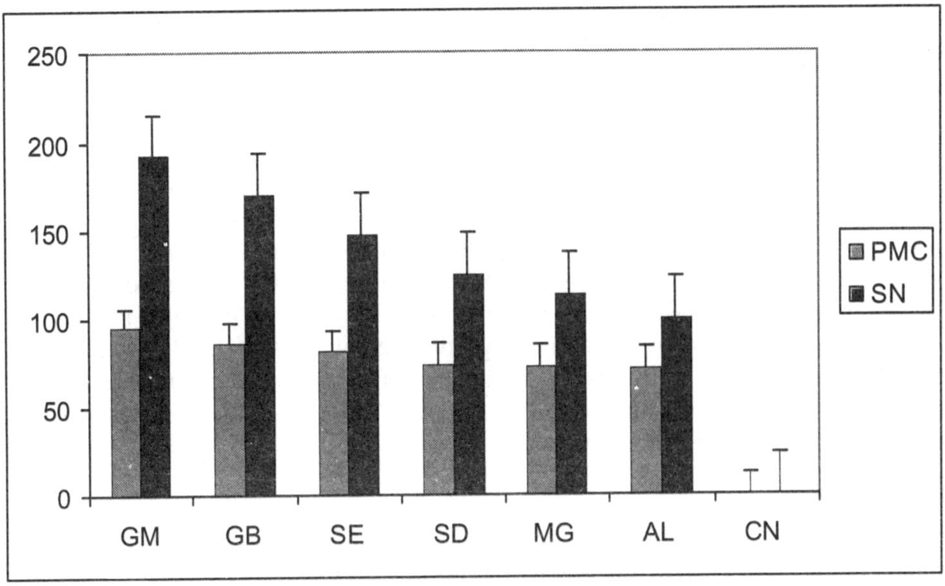

Figure 11.1: The Effect of Six AM Fungi on the per cent Root Colonization and Spore Number

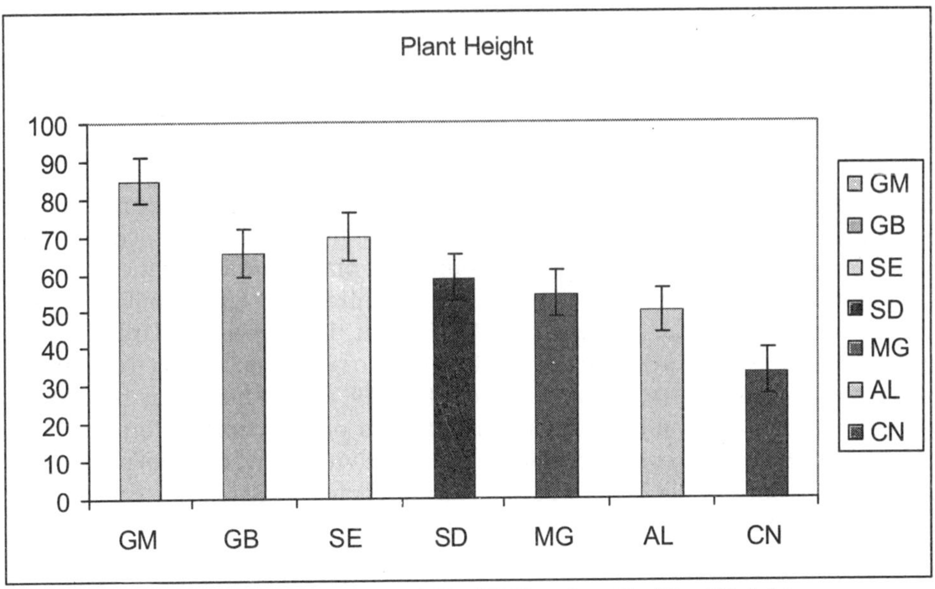

Figure 11.2: The Effect of Six AM Fungi on the Plant Height of *Hebanaria marginata* Colber

GM: *Glomus macrocarpum;* GB: *Glomus bagyarajii;* SE: *Scutellospora erythropa;* SD: *Sclerocystis dussii*; MG: *Gigaspora marginata;* AL: *Acaulospora laevis;* CN: Contro

biomass in terms of dry weight of shoot and root was measured in all the treatments and control plants. Highest increased biomass was observed in plants inoculated with *G. macrocarpum* and least was in the plants treated with *Acaulospora laevis*. Very significantly increased per cent of root colonization and spore number was recorded in mycorrhizal plants over the control plants (Figure 11.1). Experimental plants showed significantly increased per cent of mycorrhizal colonization and spore number, when they are inoculated with *G. macrocarpum* followed by the inoculation of *G. bagyarajii* and *S. erythropa* compared to remaining three AMF inoculated and control plants. Phosphorus content of shoot in all inoculated plants demonstrate that there was an increased 'P' content in shoot was observed in plants inoculated with different AM fungi over the control plant. There was significantly increased 'P' content was recorded in plants inoculated with *G. macrocarpum* compared to the other mycorrhizal plants and control plants. The data was subjected to statistical analysis by using SPSS student version.

Discussion

The present investigation clearly indicates that, the mycorrhizal inoculation enhances the plant growth and nutrient uptake by the experimental plant *Hebanaria marginata*. This increased plant growth and nutrient uptake is due to mycorrhizae, which can increase absorption of mineral nutrition; particularly immobile nutrients by host plants from the soil (Safir *et al.*, 1971). Use of mycorrhizae can increase plant biomass by enhancement of water and nutrient absorption and increasing the photosynthesis activity of plants. These results are in consistence with Mendeiros *et al.* (1994). According to Carling and Brown (1980) and Lakshman (2007), arbuscular mycorrhizal fungal species vary considerably in their efficiency to infect and influence plant growth. In the present experiment variation in increased nutrient uptake and growth of the experimental plants was observed. It is due to varying activities of individual mycorrhizae in mature orchid plants. Evidence for this was provided by Alexander and Hadley (1984) from their studies of mycorrhizas formed between *Goodyera repens* and *Rhizoctonia* sp. They found that plant growth and the uptake of Phosphorus and nitrogen were enhanced in mycorrhizal plants. The similar findings were recorded from the present investigation in the mycorrhizal plants. The effectiveness of AMF in increasing P uptake has been related to the speed and extent of root colonization by AMF (Gaur and Adholeya, 1995). AM fungi may function as a metabolic sink causing basipetal mobilization of photosynthates to roots thus, providing a stimulus for greater photosynthetic activity in mycorrhizal plants (Wu and Xia, 2006, Borkowska, 2002), due to increased photosynthetic activity inoculated plants showed better growth and nutrient uptake.

The mycorhizal plants show a greater increase in the rate of photosynthesis than their controls which may be due to increase in the content of total chlorophyll (Shrestha *et al.*, 1995; Wright *et al.*, 1998; Morte, *et al.*, 2000; Mathur and Vyas, 2000). The experimental results strongly support these findings. Plant grown in soil inoculated with *Glomus fasciculatum* showed increased chlorophyll content and phytochemical constituents (Selvaraj and Chellappan, 2006; Selvaraj, 1989), this work was in consistent with the present experimental findings. Inoculation of black gram in an unsterile soil with *Glomus epigaeum* showed increased chlorophyll and N, P and K content (Umadevi, and Sitaramaiah, 1990; Nowak, 2004). The enhancement in chlorophyll 'a', 'b' and total chlorophyll content can be attributed to the increase of absorption and translocation of essential metal ions, due to mycorrhizal infection. Analysis of the multiple interactions established by mycorrhizal fungi with plant offers new understanding of the complexity of mycorrhizas. Simplified experimental models have provided many breakthroughs on the molecular and cellular bases that regulate plant-mycorrhizal fungi interactions as well as the signaling networks active in the rhizosphere.

Conclusion

AM fungi have importance due to their capabilities on the increase in plant growth and biomass. From the results obtained from the present study, it could be concluded AM fungi have a beneficial effects on *H. marginata* grown under experimental conditions. *G. macrocarpum* caused better responses compared to all the five AM fungal species.

Acknowledgment

The authors are thankful to the UGC, New Delhi., for sanctioning Major Research Project on the "Investigation of AM fungal role in four rare millets of North Karnataka" and for financial support.

References

Alexander and Hadley (1984). The effect of mycorrhizal infection on *Goodyera repens* and its control by fungicides. *New Phytol.*, **97**: 391-400.

Bethlenfalvay, G. J. (1992). Mycorrhizae and crop productivity. In: *Mycorrhizae in Sustainable Agriculture*, (Eds.) G. J. Bethlenfalvay and R.G. Linderman. ASA Special Publication No. 54, Madison, USA, pp. 1-27.

Borkowska, B. (2002). Growth and photosynthetic activity of micropropagated strawberry plants inoculated with endomycorrhizal fungi (AMF) and growing under drought stress. *Acta Physiol Plant*, **24**: 365–370

Carling, D.E. and Brown, M.F. (1980). Relative effect of vesicular Arbuscular mycorrhizal fungi on the growth and yield of soybeans. *J. Soil. Am.* **44**: 528-532.

Davet, P. (1996). Vie microbienne du sol et production végétale. INRA, Paris.

Gaur, A and Adholeya, V. (1995). Mycorrhizal effects on the acclimatization, survival, growth and chlorophyll of micro propagated *Syngonium* and *Draceana* inoculated at weaning and hardening stages. *Mycorrhiza*, **9**: 215-219.

Gerdemann, J.W. and Nicolson, J.H. (1963). Spores of Mycorrhizal Endogone species extracted from soil. Wet sieving and decanting. *Trans. Brit. Mycol. Soc.*, **46**: 235-244.

Gianinazzi-Pearson, V. and Diem, H.G. (1982). Endomycorrhizae in the tropics. In: *Microbiology of Tropical Soil and Plant Productivity*, (Eds.) Y.R. Dommergues and H.G. Diem. Martinus Nijhoff/Dr Junk W. Publishers, The Hague.

Harley, J.L. and Smith, S.E. (1983). *Mycorrhizal Symbiosis.* Academic Press, London.

Jackson, M.L. (1973). *Soil Chemical Analysis: Advanced Course,* 2nd edn. Madison, Wisconsin, USA, p. 511.

Lakshman, H.C. (2007). Mycorrhizae: A boon for sustainable agriculture. In: *Organic Farming and Mycorrhizae in Agriculture,* (Ed.) P.C. Trivedi. IK International Publishers, New Delhi, pp. 21-35.

Mathur, N. and Vyas, A. (2000). Influence of arbuscular mycorrhizae on biomass production, nutrient uptake and physiological changes in *Ziziphus mauritiana* Lan. Under water stress. *J. Arid. Environ.,* **45**: 191-195.

Medeiros, C.A.B., Clark, R.B. and Ellis, J.R. (1994). Growth and nutrient uptake of sorghum cultivated with vesicular Arbuscular mycorrhizae isolates at varying pH. *Mycorrhiza,* **4**: 185-191.

Morte, A., Lovisola, C. and Schubert, A, (2000). Effect of drought stress on growth and water relations of the mycorrhizal association *Helianthemum almeriense* Terfezia clavery. *Mycorrhiza,* **10(3)**: 115-119.

Nowak, J. (2004). Effects of arbuscular mycorrhizal fungi and organic (eds) fertilization on growth, lowering, nutrient uptake, photosynthesis, transpiration of geranium (*Pelargonium hortorum* L.H. Bailey '*Tango Orange*'). *Symbiosis.,* **37**: 259-266.

Philips, J.M and Hayman, D.S. (1970). Improved procedure for clearing roots and staining parasite and Vesicular Arbuscular Mycorrhizal fungi for rapid assessment of infection. *Transaction of British Mycological Society.* **55**: 158-160.

Read, (1990). D.J. Mycorrhizas in ecosystems – nature's response to the 'law of the minimum'. In: *Frontiers in Mycology,* (Ed.) D.L. Hawksworth. CAB International.

Safir, G.R., Boyer, J.S. and Gerdeman, J.W. (1971). Nutrient status and mycorrhizal enhancement of water transport in soyabean. *Plant Physiol.,* **49**: 700-703.

Selvaraj, T. (1989). Studies on *Vesicular arbucular* mycorrhizae of some crop and medicinal plants. *Ph.D. Thesis,* Bharathidasan University, Tiruchirapalli, Tamil Nadu, India, p. 120.

Selvaraj, T. and Chellappan, P. (2006). Arbuscular mycorrhizae: A diverse personality. *J. Cent. Europ. Agric.,* **7**: 349-358.

Shrestha, Y.H., T. Ishll and K. Kadoya, (1995). Effect of VAM fungi on the growth, photosynthesis, of transpiration and the distribution of photosynthates of bearing Satsuma mandarin tress. *J. Ipn. Soc. Hort. Sci.,* **64**: 517-525.

Smith, S.E. and Read, D.J. (1997). *Mycorrhizal Symbiosis.* Academic Press, London, pp. 605.

Sylvia, D.M. and William, S.E. (1992). Vesicular-arbuscular mycorrhizae and environmental stress. In: *Mycorrhiza in Sustainable Agriculture,* (Eds.) G.J.Bethlenfalvay and R.G. Linderman. American Society of Agronomy Special Publication No. 54, Madison, WI, USA, pp. 101–124.

Umadevi, G. and Sitaramaiah, K. (1990). Influence of soil inoculation with endomycorrhizal fungi on growth and rhizosphere microflora on blackgram. *2nd Natl. Conf. Mycor.*, UAS, Bangalore. p. 6.

Verma, A., Sudha, Sahay, N.S., Singh, A., Kumari, M., Bharti, K., Sarbhoy, A.K., Maier, W., Walter, M., Strack, D., Franken, P., Singh, A.N. and Malla, R. (2002). *Piriformospora indica*: A plant stimulant and Pathogen inhibitor Arbuscular mycorrhizae like fungus. (Eds.) D.K. Makandey and N.R. Markandey. Capital Book Company Ltd., New Delhi, pp. 71-89.

Wright, D.P., Scholes, J.D. and Read, D.J. (1998). Effect of VAM colonization on photosynthesis and biomass production of *Trifolium repens* L. *Plant Cell Environ.* **21**: 209-216.

Wu, O.S. and Xia, R.X. (2006). Arbuscular mycorrhizal fungi influence growth, osmotic adjustment and photosynthesis of citrus under well water and water stress conditions. *J. Plant Physiol.*, **163**: 417-425.

Chapter 12

Role of Algae in Bioremediation

***Milind J. Jadhav*[1]**

ABSTRACT

Water is most important natural resource. It is a precious requirement of living world. The disposal of wastewater from domestic and industrial sources has become a serious problem throughout the world. Such disposals contaminates different water bodies and affects aquatic biota. Bioremediation by using algae, is an emerging technology for the treatment of wastewater and industrial effluents. Utilization of algae for the treatment of wastewater and industrial effluents is termed as Phycoremediation. It is most advance technology in algal biotechnology. The ultimate goal of Phycoremediation is to convert the noxious wasterwater into innocuous form. Algae utilize the waste as a nutritional source and degrade the pollutants. The advantages of Phycoremediation includes, it is cost-effective, ecofriendly and a safe process. Phycoremediation technology is implemented in three phases viz: Laboratory feasibility studies, pilot plant studies and large scale commercial plant. Pollution tolerant algae can be exploited in phycoremediation.

Keywords: Wastewater, Industrial effluents, Phycoremediation.

Introduction

Water is one of the important natural resource of earth. It is vital factor in the existence of living organisms. Rapid urbanization and industrialization polluting

---

1 Department of Botany, Sir Sayyed College, Aurangabad – 431 001, M.S., India
E-mail: dr.mjjadhav@gmail.com

water bodies at alarming rate. Existence of aquatic biota has been in danger due to disposal of wastewater and industrial effluents into rivers, ponds, lakes and seas. Toxic pollutants can be eliminated from contaminated water bodies with the help of bacteria, fungi, algae and some plants. Use of biological organisms to reduce or remove toxic pollutants from contaminated environment by degradation, assimilation or transpiration in the atmosphere is called bioremediation. Removal of organic contaminants is primarily based on either microbes naturally present at the sites or on microbial inoculants developed in laboratory and introduced at the site. Certain bacterial, fungal and algal species are capable of accumulating some toxic inorganic contaminants.

Use of Algae in Bioremediation

Algae are natural inhabitants of water bodies. They are one of the most rapid detectors of environmental pollution. Being primary producer in the aquatic ecosystem they consume a number of elements dissolved in water for their survival and growth. Polluted water harbours particular type of algae. They serves as indicators of water quality in various ways. Use of algae for the treatment of wastewater and industrial effluents is termed as Phycoremediation. Algae can remove solid, liquid and gaseous pollutants. Disinfection of sewage was also brought by Cyanobacteria through the combined effects of oxygen, hydroxyl radicals and extracellular polysaccharides (Subramanian and Uma, 2006).

Uncontrolled sewage is a big problem to health causes bad odours, epidemics and fish mortality. It is rich in organic and inorganic nutrients which under certain conditions supports a luxuriant growth of algae. Singh and Saxena (1969) studied algal succession in raw and stabilized sewage. Singh *et al.* (1970) reported fifty nine species of algae from sewage of fifteen towns of Uttar Pradesh. Blue-green algae form the major constituent of algal flora of sewage. Saha *et al.* (1985) made a survey of algal flora of habitats polluted with effluents of fertilizer factory, sugar factory, distillery and township sewage. In their survey they have recorded thirty five genera of algae belonged to Cyanophyceae, Chlorophyceae, Xanthophyceae and Bacillariophyceae. In all the polluted habitats cyanophycean members were dominant. Diatoms like *Navicula* and *Nitzschia* were also found dominant.

Treatment and disposal of sewage and industrial wastes present difficult problem. In recent years aerobic oxidation ponds are being used for treatment of sewage. Under favourable conditions oxidation ponds with algae serve as a cheap and efficient method for treatment of sewage particularly from small size communities. The work on such oxidation ponds carried out in our country has demonstrated the usefulness of this method in treatment of sewage (Purshothaman, 1958). The use of aerobic oxidation ponds for treatment of industrial waste rich in organic matter has been fully studied. A number of algal species have been isolated and studied in this connection. Technique for mass culture of *Chlorella* and *Hydrodictyon* for use in laboratory and pilot plant studies has been worked out.

Nandan and Mahajan (2003) isolated pollution tolerant cynobacteria from polluted lakes. Cyanophycean and Chlorophycean pollution tolerant algae can be used in bioremediation (Talekar, 2009 and Talekar and Jadhav, 2009). The species of

Anabaena, Arthrospira, Oscillatoria, Phormidium, Spirulina, Chlorella, Chlorococcum, Spirogyra and *Stigeoclonium* (Table 12.1) can be utilized for the removal of toxic material from polluted water bodies.

Table 12.1: Pollution Tolerant Algae which can be Used in Bioremediation

Class	*Algal Forms*
Cyanophyceae	*Anabaena sphaerica, Arthrospira platensis, Oscillatoria chlorina, O. forequi, O. obscura, O. terebriformis, Phormidium ambigum, P. bohneri, P. corium, P. mucosum, P. subincrustatum, P. usterii, Spirulina major.*
Chlorophyceae	*Chlorella vulgaris, Chloroccum humicola, Spirogyra inconstans, Stigeoclonium tenue.*

Biosorption

The ability of biological materials to accumulate heavy metals from wastewater through metabolically mediated or physico-chemical pathways of uptake known as Biosorption. It is proven to be quite effective at removing metal ions from polluted environment in a low-cost and environment friendly manner. Algae have proved to be potential metal biosorbent. Fresh water as well as marine algae have been used in biosorption of heavy metals. Aksu (2002) studied biosorption of nickel (II) ions on *Chlorella vulgaris*. Chojnacka *et al.* (2005) worked on biosorption of Cr, Cd and Cu ions by *Spirulina*.

Marine macro-algae are harvested or cultivated in many parts of the world and are therefore readily available in large quantities for the development of highly effective biosorbent materials. Kumar and Kaladharan (2006) studied biosorptive potential of *Sargassum wightii* on Cd and Pb.

Immobilized Algae in Wastewater Treatment

Nowadays researchers are utilizing immobilized algal cells for the removal of toxicant from wastewater. Rai and Mallick (1992) demonstrated a higher uptake rate for N and P by immobilized *Chlorella* and *Anabaena* than their free-living counterparts. Similar results were also observed in case of immobilized *Spirulina maxima* grown in swine waste. Tam *et al.* (1998) studied the efficiency of alginate immobilized *Chlorella vulgaris* for removing copper from solution. The amount of copper removal was related to cell densities.

Phycoremediation technology is implemented in three phases viz: Laboratory feasibility studies, pilot plant studies and large scale commercial plant. Phycoremediation is a safe, cost effective and ecofriendly technology in the field of algal biotechnology.

References

Aksu, Z. (2002). Determination of the quilibrium, Kinetic and thermodynamic parameters of the batch biosorption of nicket (II) ions onto *Chlorella vulgaris*. *Process Biotechnology*, **38**: 89-99.

Chojnacka, K.; Chojnacki, A.; Goreck, H. (2005). Biosorption of Cr3+, Cd2+ and Cu2+ ions by blue-green algae *Spirulina* sp: Kinetics, equilibrium and the mechanism of the process. *Chemosphere*, **59**: 75-84.

Kumar, V.V. and Kaladharan (2006). Biosorption of metals from contaminated water using seaweed. *Current Science* **90(9)**: 1263-1267.

Nandan, S.N. and Mahajan, S.R. (2003). Cyanobacterial diversity in polluted lakes of Jalgaon district of North Maharashtra. In: *Aquatic Environment and Toxicology*, (Ed.) Arvind Kumar, Daya Publishing House, New Delhi, p. 28-62.

Purshothaman, K. (1958). Abstract on studies on stabilisation ponds. *Bull. C.P.H.E.R.I.* **I (3)**: 88.

Raim, L.C. and Mallick N. (1992). Removal and assessment of toxicity of Cu and Fe to *Anabaena doliolum* and *Chlorella vulgaris* using free and immoblized cells. *World J. Microbiol. Biotechnol*, **8**: 110-114.

Saha, R., Saxena, P.K. and Jabeen, S. (1985). Ecological survey of the algal flora of polluted habitats of Gorakhpur. *Phykos*, **24**: 4-11.

Singh, V.P. and Saxena, P.N. (1969). Preliminary studies on algal succession in raw and stabilized sewage. *Hydrobiologia* **34**: 503-512.

Singh, V.P., Saxena, P.N., Tiwari, A. and Khan, M.A. (1970). Studies on the seasonal variation of algal flora of sewage. *Phykos* **9(2)**: 57-62.

Subramanian, G. and Uma. L. (2006). Cyanobacteria in bioremediation. ICCTABU, Plenary lecture, 8.

Talekar, Santosh (2009). Studies on algal biodiversity of Manjara river and its reservoirs in Beed district of Maharashtra. *Ph.D. Thesis*, Dr. Babasaheb Ambedkar Marathwada University, Aurangabad.

Talekar, Santosh and Jadhav, Milind (2009). Algal diversity of Polluted water reservoir. *Proc. Nat. Conf. on Biodiversity, sustainable development and human welfare*, Editor Dr. S.N. Nandan, 156-159.

Tam, N.F.Y., Wong, Y.S. and Simpson, C.G. (1988). Removal of copper by free and immoblized microalga *Chlorella vulgaris*. In: *Wastewater Treatment with Algae*, (Eds.) Y.S. Wong and N.F.Y. Tam. Springer. Verlag and Landes Bioscience, Berlin1, p. 7–35.

Chapter 13

Airborne Microorganisms Present in Government and Private Offices of Nagpur City

K.J. Cherian[1]* *and S.M. Balpande*[1]**

ABSTRACT

Microbiological study of airborne spores in the public places like Government and Private offices of Nagpur city were conducted to make awareness in the quality of air inhaled by the people. Microorganisms occurred in greater number in the air within the area of indoor environment of offices. In the present study indoor microflora (Fungi) at two locations of Nagpur city was investigated during rainy season. Few government and private offices were selected for the study. Air samples were collected by self designed air sampler for 30 minute at one place. The predominant fungi isolated from investigated air samples were *Aspergillus spp., Penicillium* spp., and *Rhizopus* spp.etc. The concentration of airborne micro flora recorded in the environment, specifically in the government offices was significantly different from private offices.

Keywords: *Airborne fungi, Air sampler, Government offices, Indoor air, Private offices.*

Introduction

The purpose of the present study was to determine the mycological contamination of atmosphere air in the Government and Private offices of the Nagpur

---

1 Department of Botany, Hislop College, Civil Lines, Nagpur, Mh., India

E-mail: *cherian_kj@yahoo.co.in, **shubhangini2009@gmail.com

city. An earlier study suggests that an indoor environment can contain a no. of filamentous fungi of which *Penicillium, Alternaria, Aspergillus, Mucor,* and *Rhizopus* etc. These fungi can release large amounts of spores, which are then carried by air streams and contaminate new environment. During breathing, people inhale these fungal spores into their bronchia and lungs and cause of many diseases and allergies (Barbara Przywitowska and Izabella Zmystowska, 2007).The aim of those studies is not only estimation of the airborne microorganisms but also their identification and the determination of factors influencing bioaerosol composition inside the offfices. Fungal flora can be hazardous for health, particularly in rooms with heating, ventilation and air-conditioning systems (Stryjakowska-Sekulska *et al.*, 2007).The present study was based on determination of airborne fungal flora during rainy season in 2010. Air samples were collected by self designed air sampler for 30 minute at one place.

Materials and Methods

Sample Collection

Samples of atmospheric air for mycological determinations were collected from five Government and Private Offices of Nagpur city. Air samples were collected by self designed air sampler, which is consisting of a conical flask, rubber cork with inlet and outlet glass tube and air motor. 20 ml Autoclaved distill water use as a collection medium in a conical flask. (Padma Srikanth and Suchitra Sudharsana, 2008). Use separate flask set for each sample. The samplers operate by drawing air through an inlet tube. Air motor was made on for 5 minute before taking each sample. 2ml of the liquid sample were inoculating on the particular Czapek's Dox media for fungi in a triplicate. After inoculation colonies of fungi are allowed to grow on Czapek's Dox culture media at 26°C temperature over 3-7 days and then identified and maintain pure culture. Slides were prepared using lactophenol and cotton blue to identify fungi.

Collection Sites

Five Government and Private Offices of Nagpur city were selected for the Studies on the micro flora of the air mainly fungi. The Government offices are SBI Nandanwan, Income Tax office, NMC Mahal office, Passport office and MOIL etc. and The Private Offices are Livepro Software Company, IT Park, Big Bazar Bardi, Goldline Phama, and OSR Technology etc.

Results and Discussion

The culture studies of fungal flora of aerosol in Government Offices were given in Table 13.1. Eight species were identified out of which *Aspergillus* spp was found in all four site while MOIL office do not contain *Aspergillus* spores. All the samples taken from Income Tax Office contain abundant quantities for spores of *Aspergillus* spp. Spores of *Penicillium* spp could be identified from passport office and MOIL Office. These two sites are located in Seminary Hill which was surrounded by thick vegetation of forest areas. Spores of *Rhizopus* spp found in NMC Mahal Office only.

The culture studies of fungal flora of aerosol in private sector offices were given in Table 13.2. Seven types of fungal spores could be identified from private sector offices. The seven spores belongs to three genera *i.e. Penicillium* spp, *Aspergillus* spp. and *Rhizopus* spp. Out of five sites the software company offices contain more fungal spores than other. The Bigbazar has thousands of people visiting every day from various parts of Nagpur city; the stalls were air-conditioned too like other sites. Still it air contains less fungal spores that too only one genus *i.e.Aspergillus* spp. The Gold line Pharama, OSR Technology offices too contain only one genus each. Only *Rhizopus* spores were found in IT Park only along with *Penicillium* spp. (Figures 13.1 and 13.2).

Table 13.1: Fungal Flora of Aerosol in Government Offices

Sl.No.	*Name of Site Govt. Offices*	*Ino Dt.*	*Sp. Dt.*	*Color of Colony*	*No. of Colony*	*Name of Fungi*
1.	SBI Nandanwan	18/09	24/09	Grey Brown Grey green	All over plate 5- 6	*Aspergillus* spp. *Aspergillus* spp.
2.	Income Tax Office	25/08	31/08	Brown, Black	All over plate	*Aspergillus* spp.
3.	NMC Mahal Office	18/08	23/08	White green Yellow Brown	All over plate 2–4	*Aspergillus* spp. *Rhizopus* spp.
4.	Passport Office	25/08	9/09	White sticky green Brown grey	Half plate 2-3	*Penicillium* spp. *Aspergillus* spp.
5.	MOIL	2/09	10/09	White Green	All over plate	*Penicillium* spp.

Table 13.2: Fungal Flora of Aerosol in of Private Offices

Sl.No.	*Name of Site Private Offices*	*Ino Dt.*	*Sp. Dt.*	*Color of Colony*	*No. of Colony*	*Name of Fungi*
1.	Livepro Software	13/09	30/09	Greenish Black	2-3 4-5	*Penicillium* spp. *Aspergillus* spp.
2.	IT Park	10/09	17/09	Yellowish Brown White mycelium Greenish, Grey White	Half plate 2-3	*Rhizopus* spp. *Penicillium* spp.
3.	Big Bazar Burdi	20/08	28/08	Greenish grey spore	All over plate	*Aspergillus* spp.
4.	Goldline Pharma	23/09	30/09	Black yellow Brown	All over plate	*Aspergillus niger*
5.	OSR Technology	16/09	23/09	Greenish	All over plate	*Penicillium* spp.

Conclusion

The fungal flora *Penicillum* and *Aspergillus* are abundant in the aerosol of indoor environment during rainy season. Majority of indoor environment tools clean and aesthetic but the air quality is found to be very poor especially in Government Offices.

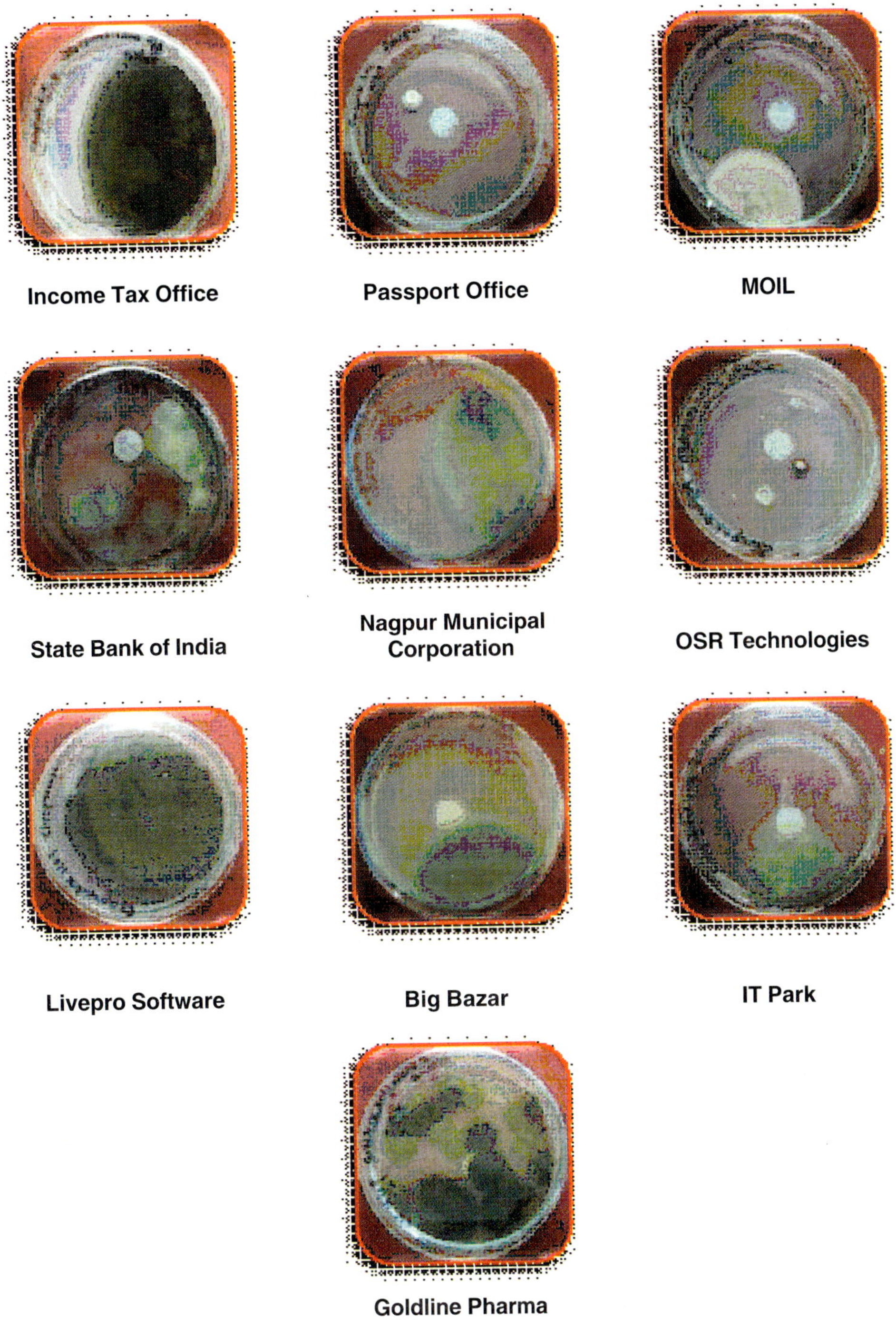

Figure 13.1: Growth of Fungal Spores at Various Sites

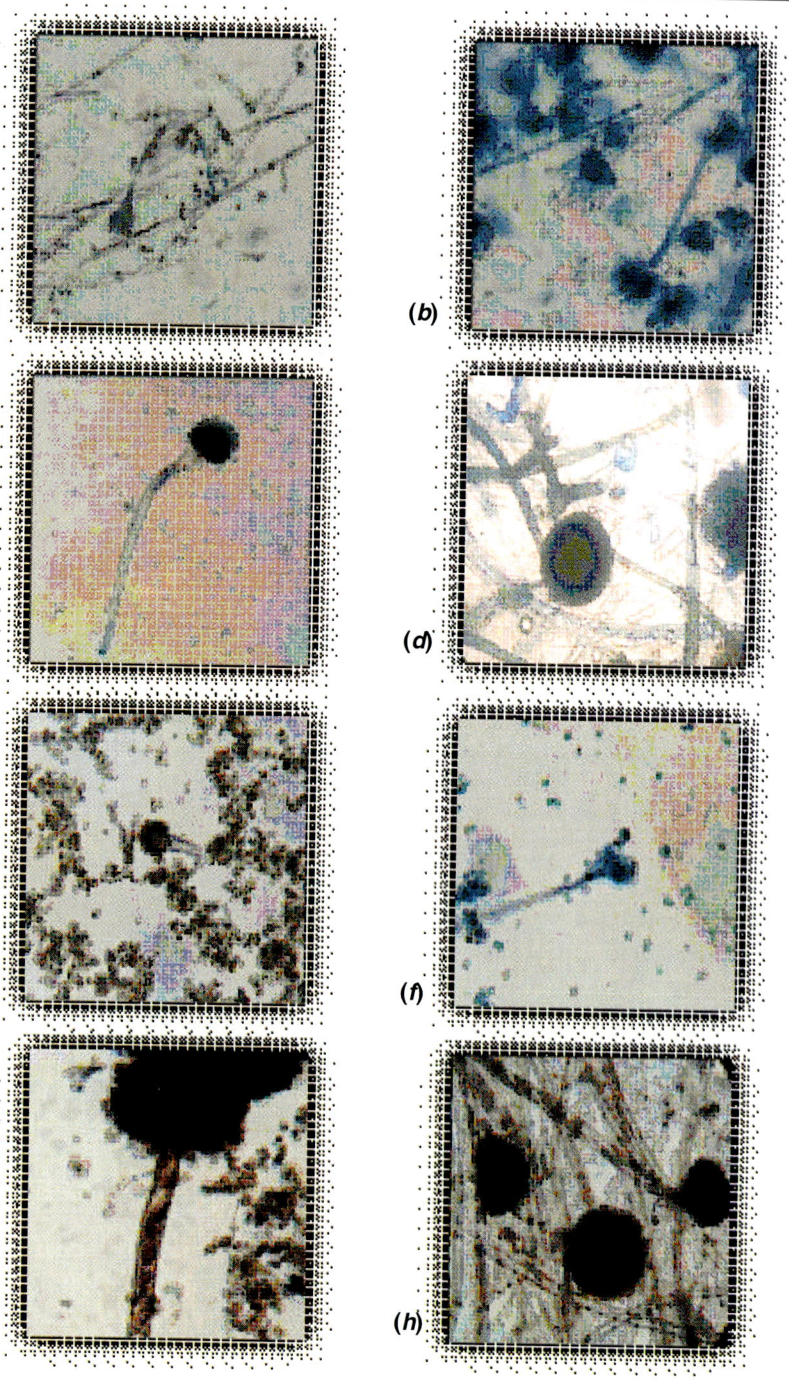

Figure 13.2: Photos Showing (*a*) *Penicillum* sp I; (*b*) *Aspergillus* sp I; (*c*) *Aspergillus* sp II; (*d*) *Rhizopus* sp; (*e*) *Aspergillus* sp III; (*f*) *Penicillum* sp II (*g*) *Aspergillus* sp IV; (*h*) *Aspergillus sp* V

References

Barbara Przywitowska, Izabella Zmystowska (2007). Quantitative Fluctuations in the fungal microflora of atmospheric air in the town of Olsztyn. *Pol. J. Natur. Sci.*, **22(3)**: 459-473.

Shrikanth, Padma and Sudharsanam, Suchitra (2008). Bio-aerosols in indoor environment: Composition, health effects and analysis. *Indian Journal of Medical Microbiology*, **26(4)**: 302-312.

Stryjakowska-Sekulska, M., Piotraszewska-Pajak, A., Szyszka, A., Nowicki, M. and Filipiak, M. (2007). Microbiological quality of indoor air in university rooms. *Polish J. of Environ. Stud.*, **16(4)**: 623-632.

Chapter 14

Interaction of Microorganisms with Mangrove Plants at Thane Coast

***Nilkanth S. Suryawanshi*[1]**

ABSTRACT

Rhizosphere, phyllosphere soil and pneumatophore microfloa of six mangrove plants growing under the sea coast of thane district was studied. On the six mangrove plants eleven species of fungi were recorded of which *Aspergillus niger* was dominant on the both dorsal and ventral surfaces. Mangroves are remarkable tropical and subtropical plants of great economic importance. A forest depletion of the mangrove has already caused soil erosion in Thane District coast so also reduce the quality of microorganism and fishery production, Hence, a large scale plantation of the mangrove in degraded coastal land is an urgently need to manage the coastal economy. The present study includes the screening of different microorganisms from soil, rhizosphere, phyllosphere, and litter of different mangrove plant namely, *Avicennia marina, Avicennia officinalis, Rhizophora apiculata, Rhizophora mucronata, Sonneratia apetla,* and *Sonneratia alba* in Thane coast. Rhizoshpere microorganisms play a vital role in mineral cycle and the ability of rhizosphere microbes to suppress the root diseases is strongly affected by environmental factors such as soil, temperature etc and its management. A total of 21 fungal species were recorded on six mangrove plants *viz. Aspergillus niger, Aspergillus flavus, Aspergillus terreus, Aspergillus humicola, Aspergillus clavatus,*

1 Research Laboratory, Department of Botany, K.V. Pendharkar College of Arts, Science and Commerce, Dombivli (E), Dist. Thane – 4212 03, Mumbai, M.S., India

Aspergillus sydowii, Fusarium moniliforme, Fusarium oxysporum, Chaetomium elongate, Chaetomium globosum, Rhizopus arrhigus, R. nigaricans, Trichoderma viride, Penicillium glomerata, P. nigricans, P. oxalic, Phoma glacts, Curvularia lunata, Cephalosporium sp. Myrothecium roridum etc. was seen in the phyllosphere of *Avicennia, Rhizophora* and *Soneratia*.

Keywords: *Mangrove plant, Microorganisms, Rhizosphere and Phyllosphere soil.*

Introduction

Interaction of microorganisms with mangrove plants is well known. There are few reports on the microorganism in the mangrove rhizosphere, soil and phyllosphere of various plants (Subharao, 1977, Gangawane, 1985, 1986, Rangaswami, 1986, Bahra, 1990,). They play very vital curial role in plant nutrition in the rhizosphere, decomposition of litter, production of phylotoalexin and growth hormone and the process of pathogenic microbes in the rhizosphere and phylloshere (Preece and Dickinson, 1971). *Rhizophera,Avicennia* and *Sonneratia* were one of the major constituents at the coastal area of Thane District. The present paper reports the microflora associated with the rhizosphere, pneumatophore, soil, phyllosphere and leaf litter of the six mangrove plants under saline condition. Protection of mangrove plant is very necessary for the microorganisms and its role.

Materials and Methods

The root, rhizosphere and phyllosphere, soil samples were collected from six mangrove plants *viz. Avicennia marina* (Forsk) Vjerh (Am), *Avicennia offcinalis* L(A0), *Rhizophora apiculata* L (Ra), *Rhizophora mucronata* Lamk (Rm), *Sonneratia apetala* Buch, Ham(Sa), *Sonneratia alba* L(Sa1), from the Thane District coast of Mumbai. For phyllosphere studies green healthy leaves where randomly selected from the above plants senescent and half decomposed leaves of these plants floating on the sea water were also selected for study of leaf litter fungi. Digging out the soil near *Rhizophora, Aviciennia* and *Sonneretia* were collected root samples, similarly soil sample (12" depth) were collected to study the micro flora. Pneumatophore samples were also collected simultaneously near the same plants. All the samples were transferred to clean and previously unused polythene bags and brought to the K.V. Pendharkar College, Research Laboratory. About 0.5 gm of the rhizosphere soil and root samples were samples were inoculated into the Pikovsky's medium. The inoculated samples were incubated at 28±2°C for six days. Phyllosphere microflora was studied by leaf print method (Balakhande, 1980). Rhizosphere and soil micro flora was analyzed by soil dilution plate technique (Timonin, 1940). Pneumatophore pieces and water wash were directly placed on the agar plate. Martins Rose Bengal Agar medium was used for fungi while Thornton's agar medium. After attaining visible growth of bacterial colonies and fungal colonies were isolated by repeated streaking and isolation in Pikovsky's agar plate and Mortin rose Bengal agar medium for fungi. Identification of fungal species was done by referring earlier literature.

Results and Discussion

A study of the rhizosphere of six mangrove plants *viz.* (*Avicennia marina* (Forsk) Vjerh, *Avicennia offcinalis* L, *Rhizophora apiculata* BL, *Rhizophora mucronata* Lamk, *Sonneratia apetala* Buch, Ham, *Sonneratia alba* L,) of Thane District coast by soil dilution plate technique were made to evaluate their micro fungal population. Both qualitatively and quantitatively the fungal flora differed considerably in the rhizosphere of all the six mangrove plants. Twenty-six fungal species were isolated from soil, root samples and leaf of above plants species. Among these fungal species the *Aspergillus niger*, was dominant. Numerous fungal species have been recorded from the rhizosphere and phyllosphere of plants growing in normal soil (Subharao Rao, 1977). In case of mangrove plants more emphasis has been given on enumeration of new species (Rovira 1965, Rai and Srivastava, 1982, Saroj, 1993, Almeida *et al.*, 1993, Nakagiri *et al.*, 1994, Tan *et al.*, 1995) recorded certain Streptomycin from the mangrove rhizosphere. Decomposition of wood and leaf litter has been studied by (Hyde and Lee, 1955).

On this background the present studies show the presence of various fungal species in rhizosphere, soil, phyllosphere of mangrove plants namely *Avicennia, Rhizophora,* and *Sonerratia* species for the first time. It is possible that these microorganisms play a significance role in the biodegradation or decomposition of litter while rhizosphere microbes help in nutrient of mangrove plants and saline environment.Very less report on fungi could colonize the pneumatophore. *Fusarium, Myrothecium, Alternaria, Phoma* and *Rhizoctonia* species are potential pathogens while *Trichoderma viride, T. harzianum* were parasitic on many fungal species and used in biological control. Fungi on leaf litter of *Avicennia, Rhizophora,* and *Sonerratia* species floating on mud and saline water were analyzed and results are shown in Tables 14.1 and 14.2. Altogether 11 fungal species were recorded on the agar plates. *Aspergillus niger* was dominant on both the surface of mangrove plant leaf and followed by *Fusarium oxysporum, Aspergillus fumigatus, Cladosporium sp., Aspergillus terrus, Aspergillus flavus, Helminthosporium, Rhizoctonia sp., Macrophomina sp., Alternaria alternate* and *Curvularia lunata.*

Quantitatively the 21 and 13 fungal colonies in the rhizosphere and non-rhizosphere of mangrove plants were recorded respectively. It was found that rhizosphere microfungi was higher in the rhizosphere than the no of fungi on non-rhizosphere (Table 14.3). Similar result was recorded by Dube and Dwivedi (1985), screened the rhizosphere and non-rhizosphere microfungi of soybean.

A total of 17 fungal species were recorded from the phyllosphere of green healthy leaves of *Avicennia* species. It was noted that 15 species were present on dorsal surface of leaves and only 13 species were recorded on ventral surface of leaves. *Aspergillus niger* was dominant on both the surface and followed by *Aspergillus flavus, A. fumigatus, Rhizopus stolonifer, Curvularia lunata, Aureobasidium oxysporium, Alternaria alternate, Cladosporium, Penicillium* sp. etc. while on ventral surface *Aspergillus fumigatus, Alternaria alternata, Aureobasidium oxysporium, Aspergillus carbonarium* were recorded *Cladosporium oxysporum, Penicillium, Penicillium funiculosum, Phoma* were not found on ventral surface while *Trichoderma harzianum* and *Phycomycetous mycelium* were

Table 14.1: Fungi Associated with Leaf Litter of *Avicennia* at Thane Coast by Leaf Print Method on Agar Plates

Sl.No.	*Fungi*	*No of Colonies/Leaf*	
		Dorsal Surface	*Ventral Surface*
1.	*Aspergillus niger*	14	07
2.	*Aspergillus flavus*	07	03
3.	*Aspergillus fumigates*	11	04
4.	*Aspergillus terreus*	09	03
5.	*Curvularia lunata*	03	02
6.	*Rhizoctonia solani*	04	02
7.	*Fusarium oxysporum*	13	07
8.	*Macrophomina phaseolina*	04	03
9.	*Cladosporium cladosporioides*	11	04
10	*Helminthosporium* sp.	05	02
11.	*Alternaria alternate*	02	—

Table 14.2: Fungi Associated with Leaf Litter of *Rhizophora* at Thane Coast by Leaf Print Method on Agar Plates

Sl.No.	*Fungi*	*No of Colonies/Leaf*	
		Dorsal Surface	*Ventral Surface*
1.	*Apergillus niger*	19	09
2.	*Aspergillus flavus*	11	09
3.	*Aspergillus fumigates*	07	03
4.	*Aspergillus terrus*	09	02
5.	*Curvularia lunata*	02	—
6.	*Rhizoctonia solani*	03	07
7.	*Fusarium oxysporum*	21	08
8.	*Macrophomina phaseolina*	15	06
9.	*Cladosporium cladosporioides*	22	10
10.	*Helminthosporium* sp.	07	03
11.	*Alternaria altenata*	03	02

not recorded on dorsal surface (Table 14.4). Table 14.5 shows that the mangrove plants namely *Avicennia marina* (Forsk) Vjerh, *Avicennia officinalis* L, *Rhizophora apiculata* BL, *Rhizophora mucronata* Lamk, *Sonneratia apetala* Buch, Ham, *Sonneratia alba* L, was recorded the maximum no of *Fusarium oxysporum, Cylendrocarpon, Apergillus niger Rhizopus nigricans* and followed by *Penicilliun, Mucor, Trichoderma viride, Botrytis, Cephalosporium, Chaetomium on Avicennia sp.* And *Rhizophora* root surface interaction of *Mucor, Phoma Botrytis* and interact with *Sonneratia* dominate fungi *Botrytis* and *Cephalosporium* etc.

Table 14.3: Availability of Some Fungi Rhizosphere and Non-Rhizosphere Soil in Mangrove Plants of Thane Creek

Sl.No.	Name of fungi	Rhizosphere	Non-Rhizosphere
1.	*Apergillus niger*	+	+
2.	*Aspergillus flavus*	+	+
3.	*Aspergillus terreus*	+	+
4.	*Aspergillus humicola*	–	+
5.	*Fusarium oxysporum*	+	+
6.	*Fusarium moniliforme*	+	+
7.	*Chaetomium globosum*	+	+
8	*Chaetomium elongate*	+	–
9.	*Myrothecium roridum*	+	+
10.	*Rhizopus nigricans*	+	–
11.	*Rhizopus arrhigus*	+	+
12	*Trichoderma viride*	–	+
13.	*Penicillium glomerata*	+	–
14	*Penicillium nigricans*	+	–
15.	*Penicillium oxalicum*	+	–
16.	*Alternaria alternata*	+	–
17.	*Cephalosporium sp.*	+	–
18.	*Curvularia lunata*	–	+
19.	*Phoma terricola*	+	+
20.	*Aspergillus clavatus*	+	–
21.	*Aspergillus sydowii*	+	+

* Triplicate replication (+ Present, and – Absent).

Table 14.4: Phyllosphere Mycoflora on Green Healthy Leaves of *Avicennia* sp. at Thane Creek by Leaf Print Method on Agar Plate

Sl.No.	Name of the Fungus	No of Colonies/Leaf	
		Dorsal Surface	Ventral Surface
1.	*Aspergillus niger*	27	07
2.	*Aspergillus flavus*	09	04
3	*Aspergillus fumigates*	12	06
4	*Aspergillus carbonarium*	04	02
5.	*Rhizophous stolonifer*	09	02
6.	*Cladosporium cladosporioides*	04	01
7.	*Cladosporium oxysporium*	03	—
8.	*Penicillium* sp.	03	—

Contd...

Table 14.4–*Contd...*

Sl.No.	Name of the Fungus	No of Colonies/Leaf	
		Dorsal Surface	Ventral Surface
9.	*Penicillium funiculosum*	02	—
10.	*Phoma*	03	—
11.	*Fusarium oxysporium*	05	02
12.	*Trichoderma harzianum*	—	01
13.	*Trichoderma viride*	03	02
14.	*Curvularia lunata*	08	02
15.	*Alternaria alternaria*	07	04
16.	*Aureobasidium oxysporium*	09	03
17.	*Phycomycetous mycelium*	-	02

Table 14.5: Microorganisms are Associated with Root Surface in Mangrove Plants of Thane Creek

Sl.No.	Name of Fungi	Am	Ao	Ra	Rm	Sa	Sa1
1.	*Penicillium glomerata*	++	++	+	++	–	–
2.	*Fusarium oxysporum*	+++	++	++	+++	–	++
3.	*Clyindrocarpon*	+++	++	+	+++	+	+
4.	*Pythium* sp.	+	++	++	++	+	++
5.	*Mucor* sp.	++	++	+++	+++	–	+
6.	*Phoma galactis*	++	+	+++	++	++	+
7.	*Aspergillus niger*	+++	++	++	++	++	++
8.	*Trichoderma viride*	++	++	++	++	+	++
9.	*Botrytis cinerea*	++	++	+++	+++	+++	+++
10	*Cephalosporium*	++	++	++	+++	+++	+++
11.	*Cheatomium globsom*	++	+	–	++	++	–
12.	*Rhizopus nigricans*	+++	++	+	++	+	+

+++: Maximum; ++: Moderate; +: Minimum; –: No Fungi

It was seen that fungal species is given in Table 14.6, eleven species were recorded from the rhizosphere while 14 species of fungal species from the soil. In the pneumatophore very few species were developed and *Phoma* was in very highest frequency no and only eight species of fungus were recorded. In the rhizosphere and soil *Cladosporium oxysporium* showed its highest percentage frequency followed by *Cladosporium cladosporioides*. On this background the present studies show the percentage of various fungal species in the rhizosphere, soil and pneumatophore of *Avicennia officinalis* for the first time. It is possible that these organisms play a very crucial role in the biodegradation of leafy litter while rhizosphere microbes help in

Table 14.6: Percentage Frequency of Fungal Species Recorded in the Rhizosphere, Soil, and Pneumatophore of *Avicennia Officinalis* L at Thane Creek

Name of the Fungal species	*Rhizosphere*	*Soil*	*Pneumatophore*
Aspergillus niger	15.45	27.12	06.45
Aspergillus flavus	12.11	09.00	04.56
Aspergillus fumigates	06.88	12.14	06.25
Aspergillus carbonarium	08.20	04.86	–
Rhizophous stolonifer	–	09.14	02.25
Cladosporium cladosporioides	35.14	44.00	01.26
Cladosporium oxysporium	52.45	56.30	–
Penicillium sp.	09.00	03.23	–
Penicillium funiculosum	–	02.45	–
Phoma sp.	05.66	06.03	–78.55
Fusarium oxysporium	08.20	12.24	–
Trichoderma harzianum	00.00	–	–
Trichoderma viride	02.00	07.36	02.23
Curvularia lunata	–	06.40	–
Alternaria alternaria	09.00	–	–
Aureobasidium oxysporium	–	09.69	03.00
Phycomycetous mycelium	–	–	–

the nutrient of mangrove under saline condition very less no of fungi could colonies the pneumatophore. *Phoma Aspergillus niger, Aspergillus flavus, Aspergillus fumigates, Rhizopus, Cladosporium Trichoderma,* are the potential to biological control of pathogenic fungi.

Acknowledgements

Financial Assistant under Minor Research Project by UGC (WRO) Pune, is gratefully acknowledged. I also specially thanks to Prof. L.V. Gangawane, Dr. Babasaheb Ambedkar Marathwada University, Aurangabad, for suggestion and motivate to this work.

References

Almeida, F., Marques, O., Bueno, R. and Bononi, V.L.R. (1993). Some basidiomycetes from mangrove in Sao Paulo State. *Hoehnea*, **20**: 87-92.

Bahra, A. (1990). Ecological studies on rhizosphere mycoflora of some economic plants of Rajasthan. *Acta. Ecol.*, **12(2)**: 123-129.

Balakhande, L.D. (1980). Studies on phylloplane mycoflora of crop of Marathwada. *Ph D. Thesis,* Marathawada University of Aurangabad, India.

Gangawne, L.V. (1985). Crop management and rhizosphere mycoflora: A review. *Ind. Bot. Reptr.*, **4**: 160-168.

Gangawane, L.V. (1986). Phylloshere microflora in relation to leaf protein extraction. *Ind. Bot. Reptr.*, **4**: 144-147.

Hyde, K.D. and Lee, S.Y. (1955). Ecology of mangrove fungi and our knowledge? *Hydrobiologia,* **295**: 107-118.

Nakagiri, A., Steven, Y.N. and Iadayashi, I. (1994). Two new *Halophytlphthora* spp. and *H.tartarea* and *H. tasteri* from intertidal decomposing leaves in salt marsh and mangrove regions. *Mycoscience.*, **35**: 223-232.

Preece, T.F. and Dickissin, C.H. (1971). *Ecology of Leaf Surface Microorganisms* (Ed.) Acad Press London.

Rangawami, G. (1986). Soil plant microbe interrelationship. *Ind. Phytopath.*, **41(2)**: 165-172.

Rovira, A.B. (1965). Plant root exudates and their influence upon soil microorganisms. In: *Ecology of Soil Borne Plant Pathogen*, pp. 170-186.

Rai, B. and Srivastava, A. (1982). Decompostion of leaf litter in relation to microbial population and their activity in tropical dry mixed deciduous forest. *Pedobio.*, **24**: 151-159.

Saroj, S. (1993). Fungal flora of rhizosphere soil in arid and semiarid regions associated with soil fertility. *Proc. Sci. Cong.*, p. 34.

Subha Rao, N.S. (1977). *Soil Microorganisms and Plant Growth*. Oxford and IBH publishing Co., New Delhi, pp. 226-239.

Tan, T.K., Teg, C.J. and Jones, E.B.G. (1995). Substrate type and microbial interaction as factors affecting ascocarp formation by mangrove fungi. *Hydrobiologia,* **295**: 127- 134.

Timonin, M.I. (1940). The interaction of higher plants and soil microorganism. I. Microbial population of the rhizosphere of seedling of certain cultivated plants. *Canad J. Res.*, pp. 307-317.

Chapter 15

Arbuscular Mycorrhizal Association in Some Medicinal Plants Belonging to Asteraceae Found in Panhala Hill of Maharashtra, India

Vishal R. Kamble[1*], *Rupesh R. Maurya*[1] *and Dinesh G. Agre*[2]

ABSTRACT

Arbuscular mycorrhizal (AM) fungi normally colonizes almost all tropical plants, however the incidence of mycotrophy in medicinal plants is less documented compared to studies on forestry and crop species. The present study was undertaken to evaluate the association of AM fungi in some medicinal plants, belonging to Family Asteraceae from Panhala Hill and fort region in Kolhapur District located in Western Ghats of Maharashtra. The degree of AM colonization varied in all the 10 plant species. Maximum colonization was recorded in *Ageratum conyzoides* (96 per cent) followed by *Gynura nitida* (87.33 per cent),

1 Department of Botany, Bhavan's College Andheri (West), Mumbai – 400 058, Mh., India

2 Department of Biology, Utkarsha Vidyalaya and Jr. College, Virar (W) – 401 303, Mh., India

* Corresponding Author E-mail: vrksiddhant@rediffmail.com

Siegesbeckia orientalis (86.95 per cent), *Tricholepis amplexicaulis* (86.67 per cent) and *Vernonia cinerea* (80 per cent). AM fungal spores belonging to *Acaulospora, Glomus* and *Scutellospora* were isolated from rhizosphere soil with *Glomus* being the dominant Genus.

Keywords: AM fungi, AM colonization, Medicinal plants, Asteraceae.

Introduction

The demand for medicinal plants is increasing with the rise in the consumption of herbal in India. There is steady increase in the cultivation of medicinal plants to maintain a steady supply to support the increasing demand. Since last few decades the use of herbal to ameliorate human sufferings are importance at par with the synthetic drugs which has inspired the Government of India and plant scientists to promote the conservation of natural flora including medicinal plants and their demand of herbals (Roy *et al.*, 2007). Arbuscular mycorrhizal (AM) fungi normally colonizes almost all tropical plants, however the incidence of mycotrophy in medicinal plants is less documented compared to studies on forestry and crop species. Since AM association in medicinal plants has received little attention as compared to studies on some economically important species and crops. The purpose of present study is to investigate the extent of AM association in medicinal plants from Panhala which is located at 16°49′N 74°07′E16.82°N 74.12°E. It is known for its historical importance. The study area is situated at an altitude of 3100 ft.18 km northwest of Kolhapur, in Kolhapur district of Maharashtra. In present study an evaluation of AM fungi associated with roots of 10 medicinal plants *viz., Ageratum conyzoides* L, *Eupatorium adenophorum* Spreng, *Flaveria trinervia* (Spr) C., *Gnaphalium polycaulon* Pers., *Gynura nitida* DC., *Siegesbeckia orientalis* L., *Sphaeranthus indicus* L., *Tricholepis amplexicaulis* Cl., *Tridax procumbens* L. and *Vernonia cinerea* (L) Less belonging to the Family Asteraceae was carried out.

Materials and Methods

Root Colonization

The roots and rhizosphere soil samples of all the medicinal plants undertaken for the study were collected from study area during Nov.–Dec. 2010. Phillips and Hayman (1970) procedure was employed to clear and stain the non-pigmented roots, while the pigmented roots were treated following the method of Kormanik *et al.* (1980). A species was considered mycorrhizal if the root samples showed hyphae, vesicles or arbuscules (Pendleton and Smith, 1983; Pond *et al.*, 1984). Percent AM colonization was calculated by the formula of Giovannetii and Mosse (1980).

Spore Types and Quantification

AM spore types and their quantification in soil samples were recorded after extraction by the wet sieving and decanting technique (Gerdemann and Nicolson, 1963). Spores were mounted in Polyvinyl-lactoglycerol (PVLG), examined for their various morphological characters and identified by using the literature and manuals (Hall, 1980; Hall and Fish, 1979; Pacioni, 1992; Schenk and Perez, 1990).

Results and Discussion

The study revealed that the degree of AM formation varied in all plant species studied (Table 15.1). Five of the 10 plant species *viz., A. conyzoides, G. nitida, S. orientalis, T. amplexicaulis* and *V. cinerea* recorded more than 70 per cent colonization while in the remaining five species *viz., E. adenophorum, F. trinervia, G. polycaulon, S. indicus* and *T. procumbens* the colonization levels ranged from 25-50 per cent of the total root length. Gupta and Mukerji (2006) reported the absence of AM colonization in *Gnaphalium purpurium*, while our study reports 28.57 per cent colonization in the related species *viz., G. polycaulon*. Hafeel and Gunatilleke (1988) reported AM colonization in *Eupatorium odoratum* found growing only in *Pinus* plantations, while did not find colonization in plants growing in natural forest and low land rainforest of Sri Lanka. The variation in root colonization may be attributed to variations in environmental factors.

Table 15.1: Per cent Root Colonization and Spore Density in Medicinal Plants Belonging to Asteraceae

Sl.No.	Plant Species	Colonization		Colonization (%)	Spore Density/ 100g
		Vesicles	Arbuscules		
1.	*Ageratum conyzoides* L	++++	–	96.00	391
2.	*Eupatorium adenophorum*	++	–	35.00	191
3.	*Flaveria trinervia*	++	+	50.00	93
4.	*Gnaphalium polycaulon*	++	–	28.57	104
5.	*Gynura nitida*	+++	–	73.33	46
6.	*Siegesbeckia orientalis*	++++	–	86.95	25
7.	*Sphaeranthus indicus*	++	++	43.75	123
8.	*Tricholepis amplexicaulis*	++++	+	86.67	213
9.	*Tridax procumbens*	++	–	48.64	35
10.	*Vernonia cinerea*	++++	–	80.00	77

Legend: 1-25 per cent = +; 25-50 per cent = ++; 50-75 per cent = +++; > 75 per cent = ++++–= absent.

All values are mean of 5 replicates.

Highest spore count was reported in *A. conyzoides* and least in *S. orientalis* (Table 15.1). However, it is very difficult to make any conclusion on the basis of comparison between colonization percentage and spore density. AM fungal spores belonging to three genera *viz., Acaulospora, Glomus and Scutellospora* were recovered from rhizosphere soils. The genus *Glomus* was dominant. However, further studies are needed to evaluate species diversity of AM fungi. Mycorrhizal fungi play an intricate role in the formation of soil aggregates in association with other soil microbes (Miller, 1987) and also in plant growth and yield (Thapar and Khan, 1985; Kormanik, 1985). Although the plants under study are having medicinal potential, but they are not yet brought under commercial cultivation and some of them are considered as weeds (Gupta and Mukerji, 2006). It is well known that to introduce any cultivation practice for wild medicinal plant species it is very important to study natural association of microorganisms like AM fungi and hence, in present context the results obtained from this study will be useful.

Acknowledgements

The authors would like to thank Dr. R. M. Mulani, S. R. T. M. University, Nanded for his valuable suggestions. Thanks are also to due to Principal, Dr. V. I. Katchi, Dr. M. S.Chemburkar and Prof. Sriharsha Head of Botany Department Bhavan's College, for providing laboratory facilities.

References

Gerdemann, J.W. and Nicolson, T.H. (1963). Spores of mycorrhizal *Endogone* species extracted from soil by wet sieving and decanting. *Trans. Br. Mycol. Soc.*, **46**: 235-244.

Giovannetii, M. and Mosse, B. (1980). An evaluation of techniques for measuring vesicular-arbuscular mycorrhizal infection in roots *New Phytol.*, **84**: 489–500.

Gupta, R. and Mukerji, K.G. (2006). Vesicular arbuscular mycorrhiza in some weeds belonging to compositae and gramineae. In: *Mycorrhiza*, (Eds.) A. Prakash, V.S. Mehrotra, Scientific Publication. India, 161-163.

Hafeel, K.M. and Gunatilleke, I.A.U.N. (1988). Distribution of endomycorrhizal spores in disturbed site of low land rainforest in Sri Lanka. In: *Procedings of Mycorrhizae for Green Asia at First Asian Conference on Mycorrhizae*, Madras University, India, p. 37-45.

Hall, I.R. (1980). Growth of *Lotus pedunculatus* Cav. in an eroded soil containing soil pellets infected with endomycorrhizal fungi. *N. Z. J. Agri. Res.*, **23**: 103–105.

Hall, I.R. and Fish, B.J. (1979). A key to the endogonaceae. *Trans. Br. Mycol. Soc.*, **73**: 261–270.

Katoch, R., Sharma, O.P., Dawra R.K. and Kurade, N.P. (2000). Hepatotoxicity of *Eupatorium adenophorum* to rats. *Toxicon.*, **38(2)**: 2309-314.

Kormanik, P.P. (1985). Effect of phosphorus and vesicular arbuscular mycorrhizae on growth and leaf retention of black walnut seedlings. *Can. J. Tree. Res.*, **15**: 688-693.

Kormanik, P.P., Bryan, W.C. and Shultz, R.C. (1980). Procedure and equipment for staining large number of plant root samples for endomycorrhiza. *Can. J. Microbiol.*, **26**: 536-538.

Miller, R.M. (1987). The ecology of vesicular arbuscular mycorrhizae in grasslands and shrublands. In: *Ecophysiol*, (Eds.) VAM Pl., G.R. Safir. CRC Press, Boca Raton, Finland, pp. 135-170.

Pacioni, G. (1992). Wet-sieving and decanting technique for the extraction of spores of vesicular arbuscular mycorrhizal fungi. *Meth. Microbiol.*, **22**: 317–322.

Pendleton, R.L. and Smith (1983). *Oecologia*, **59**: 296-301.

Phillips, J.M. and Hayman, D. S. (1970). Improved procedure for clearing roots and staining parasitic and vesicular arbuscular mycorrhizal fungi for rapid assessment of infection. *Trans. Br. Mycol. Soc.*, **55**: 158–161.

Pond, E.C., Menge, J.A. and Jarrel, W.M. (1984). *Mycologia*, (76): 74-84.

Roy, A.K., Singh, A.N. and Gautam, N.K. (2007). Rhizosphere AM fungi of some rare medicinal plants. In: *Rhizosphere Biotechnology: Plant Growth Retrospect and Prospect*, (Eds.) A.K. Roy, B.N. Chackraborty, D.S. Mukadam and Rashmi. Scientific Publisher India, pp. 201-206.

Schenk, N.C. and Perez, Y. (1990). *Manual for the Identification of VA-Mycorrhizal Fungi*, 3rd Edn. University of Florida, Gainesville, Florida, pp. 249.

Thapar, H.S. and Khan, S.N. (1985). Effect of VA-mycorrhiza on the growth of hoop pine. *J. Tree. Sci.*, **4**: 39-43.

Chapter 16

Native AM Fungi in Agroforestry Tree Species with Seasonal Variation

Krishna H. Waddar[1] *and H.C. Lakshman*[1]

ABSTRACT

The plant rhizosphere represents highly complex ecosystem, which is influenced by number of biotic and abiotic factors. Mycorrhiza in the rhizosphere has strong influence on plant growth and health. In the present study four agroforestry trees *viz., Pongamia pinnata* L., *Azadirchta indica* A Juss., *Bauhinia variegate* L., and *Tamarindus indica* L. were selected. Roots and soil samples were collected form where these plants growing in agroforestry. Studies on seasonal changes of AM fungal spore population and per cent of root colonization was determined at the end of each growing seasons such as summer, winter, and rainy seasons. Results revealed that, higher spore number and moderately root colonization was observed in three species such as *Pongamia pinnata* L., *Azadirachta indica* A Juss. and *Bauhinia variegata* L. during summer. But in *Tamarindus indica* L. rhizosphere exhibited higher spore number and significantly increase in per cent of colonization in summer seasons. However, per cent of root colonization and spore number no significant improvement was observed during winter seasons. Mycorrhizal porpagules shows moderate to poor in ten different sites. *Glomus* was dominant AM fungi in all the examined sites followed by *Gigaspora,*

---

1 Microbiology Lab, P.G. Department of Studies in Botany Karnatak University, Darwad – 580 003. Karnataka, India

Acaulospora, Scutellospora and *Sclerocystis* respectively. The ecology of soil and seasonal changes in spore and colonization in three seasons and have been discussed.

Keywords: *Seasonal changes, Agroforestry trees, Root colonization and Spore number.*

Introduction

In Agroforestry, a land use system and technology in which trees are deliberately planted on the same land with agricultural crop and/or animals have been recognized as one of the most promising strategy for rehabitating the already degraded areas. The benefit of agroforestry includes the amelioration of soil chemical and physical properties to induction of soil erosion, improved weed control and increased availability of fuel wood and/or fodder (Young 1997; Chin Ong and P. Huxley, 1996). The degree to which an agroforestry system can provide the above benefits particularly depends on the quantity of biomass on agroforestry tree species can produce.

The potential benefit of mycorrhizal fungi in rehabilitation of degraded areas by use of agroforestry system is more apparent than ever before. The need to increase food, fire and fuel wood production to keep pace with fast growing population in developing countries in the tropics is crucial. The low biomass production of agroforestry tree species in degraded areas can therefore be circumvented by the use of mycorrhizal fungi in an agroforestry (Lakshman 2008). Agronomist and other have drawn attention to the serious and mounting ecological problems caused by deforestation and loss of tree on farm lands and water sheds. Despite short coming in the available evidences, it is widely agreed that large and increased losses occur each year because of agriculture and clearing commercial fodder and building materials. In many respects, the loss of trees is not surprising given population growth and the increasing demands for arable land, fuel wood and timber.

Since three decades AM fungi has gained prominence in view of their role in improving nutrient uptake, per cent survival, growth performance and biomass yield of plant species (Reena and Bhagyraj, 1999; Rachel *et al.*, 1993; Durga and Gupta, 1995).Very little work has been carried out on the mycorrhizal association of forest trees (Rahangdale and Gupta, 1999) an indigenous and introduced flora of humid tropical forest in particular. The present investigation on important tree species from dry deciduous forest of Dharwad district for exploiting them in energy/biomass plantation programme.

During recent times an enormous amount of research has been done on the potential role of mycorrhizal in the revegetation of disturbed landscapes (Allen and Allen, 1980; Draf and Nicolson, 1974; Jasper *et al.*, 1987; Reeves *et al.*, 1979; Stathi *et al.*, 1988; Zak *et al.*, 1982). Arbuscular Mycorrhizal (AM) fungi are now well practiced in forestry management (Mukerji *et al.*, 1996). Most tree species requires a symbiotic association with microorganisms for successful establishment and growth on hospitable sites (Jorgensen 1979). They help in establishment of the forest tree species by increasing the volume of soil exploration and uptake of nutrients, biocontrol and plant diseases and growth of plant species used in afforestation programmers (Bakshi, 1974; Byrareddy and Bagyaraj, 1988; Thapar and Khan, 1973). Survey of Arbuscular mycorrhizal fungi in different tree species were studied in India (Kumar *et al.*, 2000). As such, the present study was carried out to observe the Arbuscular mycorrhizal infection and spore number on roots of some agroforestry tree species.

Material and Methods

A survey was made to examine AM fungi formation in natural dense population of agroforestry trees species where they are growing. The commonly growing tree species include 63 (sixty three) genera belong to various families. The soil was taken randomly chosen from five selected Agroforestry location at Dharwad district, Karnatak, which is geographically located between 14° 15′ and 15° 50′ north longitude and 74° 48′ and 76° 20′ east latitude. Roots and rhizospheric soil samples were collected form where these plants growing in agroforestry and season wise. The root samples were washed thoroughly in tap water and cut into 1cm fragments and autoclaved in 30 minutes in 10 per cent KOH solution and cleaned in distilled water and neutralized in 1 per cent HCL and stained 0.05 per cent Trypan blue in lactophenol. Per cent of AM colonization were estimated following the slide method (Philips and Hayman, 1970). The spore were separated from the soil by wet sieving and decanting technique (Gerdemann and Nicolson,1963).Recovered spores were identified with the help of manual and different taxonomic key propose by (Schenck and Perze, 1990).

$$\% \text{ of Colonization} = \frac{\text{Number of segments colonized}}{\text{Total number of root segments examined}} \times 100$$

Results

Root samples of all the tree species examined from the five location at Dharwad district, Karnataka. Experimental results showed in (Table 16.1) that, all the sixty three indigenous tree species were associated Arbuscular mycorrhizal (AM) fungi. Though the percent of root colonization and spore number was varied. Total AM colonization in different tree species varied significantly as indicated. The range of colonization varied from 22 per cent to 91.5 per cent. The highest AM colonization was recorded *Azadirchta indica* A Juss (91.5 per cent) and the lowest was recorded in *Erythrina indica* Lamk (22 per cent). Results revealed that, higher spore number and moderately per cent root colonization was observed in three species such as *Pongamia pinnata* L., *Azadirachta indica* A Juss. and *Bauhinia variegata* L. during summer. In *Tamarindus indica* L. rhizosphere exhibited higher spore number and significantly increased per cent of colonization in winter seasons. However, per cent of root colonization and spore number not increased during rainy seasons showed in (Figures 16.1–16.4). The results of the soil assessment of different tree species have been recovered. The total spore number of AM fungi in the rhizospheric soils of different all 63 tree species varied significantly as indicated. The highest spore number 50gm/soil was recorded in the rhizosphere soil of *Terminalia chebula* L. (137) and the lowest spore number 50gm/soil was recorded in *Lagerstromia indica* L (29).

Mycorrhizal porpagules showed not much improved in all the five different sites. *Glomus* was dominant AM fungi in most of the examined sites followed by *Gigaspora, Acaulospora, Scutellospora* and *Sclerocystis* respectively. This reflects the mycotropic nature of all plant species distributed and ability of AM fungi in soil to associated wide range of host species.

Table 16.1: Effective Higher, Moderate and Lower Colonization of Arbuscular Mycorrhizal Fungi in the Agroforestry Tree Species

Sl.No.	*Tree Species*	*Per cent of Colonization MISE*	*Spore Number per 100 gm Soil*
1.	*Feronia elephantum* Correa.	86.5±6.3	114
2.	*Murray paniculata* Jack.	90.2±3.5	118
3.	*Azadirachta indica* A Juss.	89.5±7.1	103
4.	*Acacia farnesiana* Willd.	91.5±5.1	100
5.	*Bauhinia variegata* L.	79.5±3.3	128
6.	*Tamarindus indica* L.	72.5±5.1	111
7.	*Terminalia arjuna* Bedd.	88.7±2.7	136
8.	*Terminalia belerica* Roxb.	86.5±6.4	122
9.	*Terminalia chebula* Retz.	87.3±4.5	137
10.	*Tectona grandis* L.	80.5±4.0	128
11.	*Anogeissus latifolia* Wall.	76.5±5.5	113
12.	*Cinnarnum zeylanica* blame.	77.5±3.1	135
13.	*Casuarina equisetifolia* forest.	80.3±2.1	78
14.	*Lagerstromia indica* L.	90.4±1.1	84
15.	*Samanea saman* Merr.	89.5±6.3	108
16.	*Swietania macrophylla* king.	80.7±5.5	102
17.	*Michelia champaca* L.	77.2±8.0	78.0
18.	*Polyanthia longifolia* Thw.	72.5±6.5	83.2
19.	*Garcinia indica* chois.	49.5±3.4	71.0
20.	*Shorea robusta* Gaertn.	52.5±2.1	104
21.	*Citrus maxima* Merr.	68.3±1.1	62
22.	*Semecarpus anacardiam* L.	70.8±2.4	77
23.	*Acacia leucophluea* willd.	67.5±3.6	108
24.	*Bauhinia purpurea* L.	66.3±5.2	103
25.	*Pongamia pinnata* (L) perrie.	71.3±4.3	59
26.	*Pterocarpus marsupium* Roxb.	68.5±6.0	57
27.	*Syzggium cumini* Skeels.	72.8±3.2	64
28.	*Adina cordifolia* L.	72.5±1.2	44
29.	*Mitragyna parvifolia* Korth.	70.2±1.8	105
30.	*Madhuca longifolia* L.	69.8±2.0	115

Contd...

Table 16.1–*Contd...*

Sl.No.	Tree Species	Per cent of Colonization MISE	Spore Number per 100 gm Soil
31.	*Diospyros assimilis* Bedd.	66.5±3.4	88
32.	*Gymnema sylveste* R. Br.	78.3±5.2	79
33.	*Grevillea robusta* A. Cun.	70.2±5.5	102
34.	*Trema orientails* Blum.	68.3±3.1	113
35.	*Eucalyptus lanceolatus* L.	61.5±2.1	79
36.	*Anacardia occidentale* L.	72.5±1.1	62
37.	*Garcinia purpurea* Roxb.	74.5±5.4	68
38.	*Careya arborea* Roxb.	50.5±8.3	70
39.	*Cassia fistula* L.	54.5±4.2	92
40.	*Lagerstromia hypolenca* Kurtz.	55.8±3.4	62
41.	*Ziziphus jujuba* Lamk.	40.8±2.7	68
42.	*Mangifera indica* L.	39.5±5.2	72
43.	*Acacia arbica* willd.	43.5±3.3	105
44.	*Albizzia lebbeck* Benth.	29.8±3.5	102
45.	*Hardwickia binata* Roxb.	32.8±4.3	92
46.	*Peltophorum pterocarpum* Backer.	34.5±6.4	100
47.	*Xylia xylocarpum* Roxb.	39.5±6.4	111
48.	*Terminalia paniculata* Roth.	22.5±7.2	99
49.	*Lagerstromia speciosa* Pers.	28.9±8.1	79
50.	*Lawsonia alba* Lamk	42.5±2.8	64
51.	*Madhuca indica* Gmell.	34.2±5.1	52
52.	*Mimusops elengi* L.	28.2±4.4	58
53.	*Solanum xanthocarpum* S&W.	32.8±4.3	78
54.	*Artocarpus hirsute* L.	48.2±4.5	72
55.	*Artocarpus integrifolia* L.	42.9±6.1	109
56.	*Terminalia arjuna* Bedd.	41.1±4.3	74
57.	*Glircidia machlata* L.	38.8±8.2	54
58.	*Erythrina indica* Lam.	38.8±2.1	80
59.	*Gravelia arborea* L.	22.3±3.6	102
60.	*Premna integrifolia* L.	36.5±2.7	59
61.	*Santolum album* L.	22.8±7.1	38
62.	*Phyllanthus emblica* L.	38.9±3.2	34
63.	*Lagerstromia indica* L.	39.8±4.1	29

Table 16.2: Seasonal Changes in the Number of AM Spores per 50 gm Soil in Experimental Plants of Dharwad District

Season	*Soil Depth (cm)*	*Pongamia pinnata (L) Pierre*	*Azadirchta indica L*	*Bauhinia varigata L*	*Tamarindus Indica L*
Summer	0-10	381±1.3	412±23	402±2.1	251±1.5
	10-20	203±2.1	266±2.8	224±2.1	169±0.5
	20-30	86±0.5	104±1.4	92±2.0	73±0.5
	30-40	29±1.0	37±1.2	2 1±1.4	24±1.5
Rainy	0-10	203±1.1	234±1.2	251±1.0	280±1.0
	10-20	137±1.4	111±0.2	173±1.0	160±1.5
	20-30	69±2.2	67±1.5	88±1.2	28±1.2
	30-40	44±0.5	48±2.0	54±1.4	39±1.5
Winter	0-10	372±1.2	410±1.4	386±0.2	400±1.6
	10-20	261±1.8	242±2.0	194±0.5	132±1.0
	20-30	727±2.1	98±2.1	77±2.1	101±2.0
	30-40	30±2.1	35±1.1	27±1.2	31±2.1

Discussion

Overall, we recovered, all sixty three tree species examined from all the five location at Dharwad district, Karnataka, and it reflects the ambiguous and well developed mycorrhizal association. This is a surprisingly large number, given that so far only about 150 AMF species have been described worldwide for the phylum *Glomeromycota*. Arbuscular mycorrhizal (AM) fungal colonization and spore population varied significantly in different tree species. The variation in the percent root colonization and AM fungal spore number recorded. In the present study might explain that the plant species with regard to AM fungal spores and colonization had a narrow to broad range. The intensity of AM fungal association and spore population shows great seasonal fluctuations, during the different seasons of the year. In all five localities, forty one species of AM fungal species recorded. The results were in consistent with the reports of earlier workers (Muthukumar and Udayan, 2000; Ragupathy and Mahadevan, 1993; Sharma *et al.*, 1986). Ongeune and Kuyper (2001) reported the variation in AM colonization in different trees from rain forest of south Cameroon. Assay of rhizosphere soils from sixty three tree species for AM fungal spore revealed a high spore count of AM fungi belonging to 4 genera *Glomus, Gigaspora, Acaulospora* and *Scuttilospora*. Among which *Glomus* was found to be the predominant in all the soil samples similar observation was made by (Sharma *et al.*, 1986: Lakshman, 1996, 2008). *Glomus* was the most predominant spore genera in naturally growing in rhizosphere of tree species.

The population of AM fungal spore was occur very high during early summer and in all the tree species except in *T. indica* L. from all five localities, being highest during the late winter season in *T. indica* L. Spore population was suppressed during early winter and rainy season in all the samples observed. Intensity of AM fungal

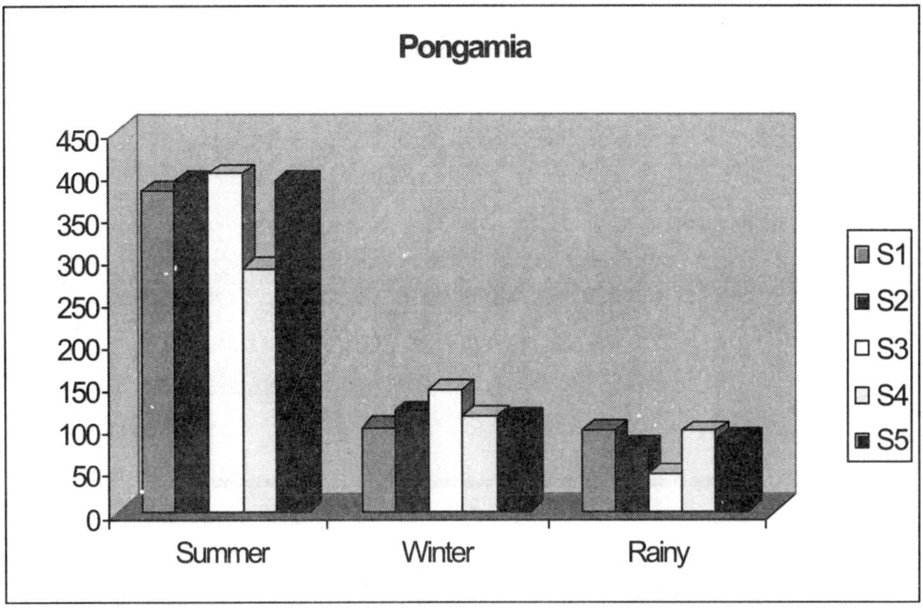

S1: Amminabhavi; S2: Manasur; S3: Mugad; S4: Navalur; S5: Kyarakoppa

Figure 16.1: Effect of Different Seasons on the Occurrence of AM Fungal Spores in Soils of *Pongamia pinnata* (L) Pierre

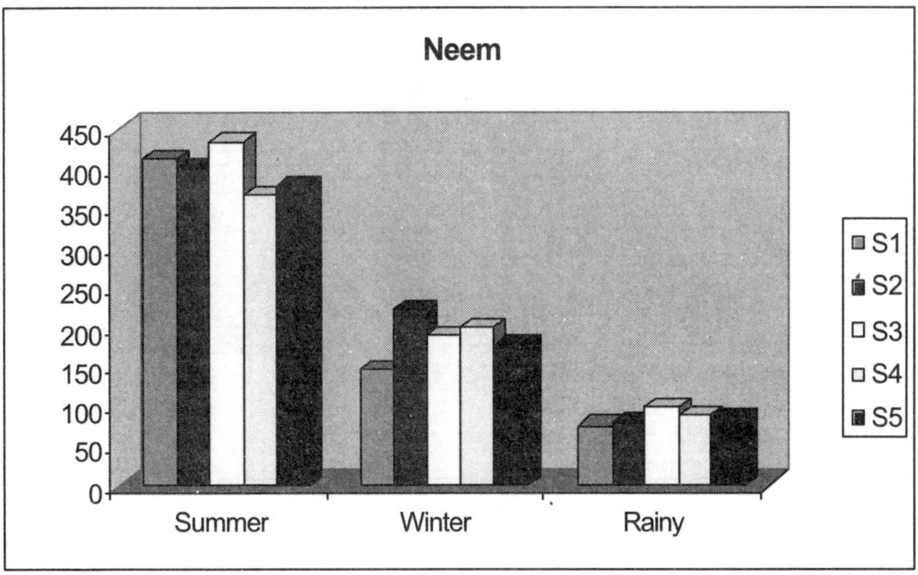

S1: Amminabhavi; S2: Manasur; S3: Mugad; S4: Navalur; S5: Kyarakoppa

Figure 16.2: Effect of Different Seasons on the Occurrence of AM Fungal Spores in Soils of *Azardirchta indica* L.

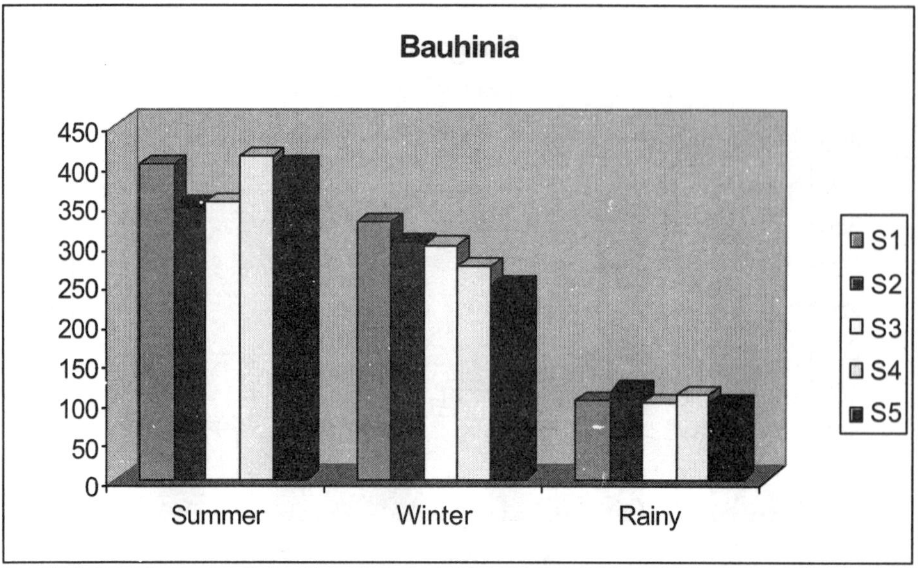

S1: Amminabhavi; S2: Manasur; S3: Mugad; S4: Navalur; S5: Kyarakoppa

Figure 16.3: Effect of Different Seasons on the Occurrence of AM Fungal Spores in Soils of *Bauhinia variegata* L.

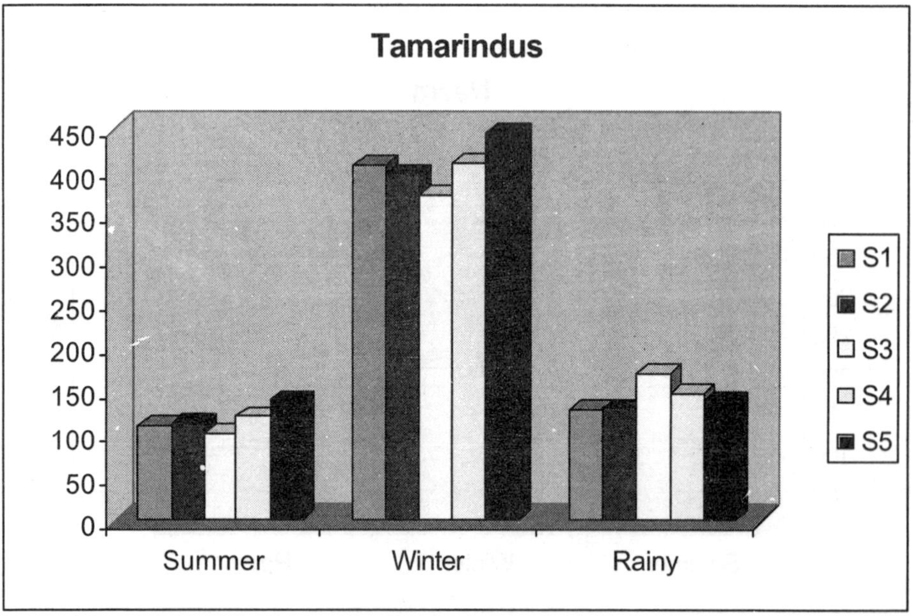

S1: Amminabhavi; S2: Manasur; S3: Mugad; S4: Navalur; S5: Kyarakoppa

Figure 16.4: Effect of Different Seasons on the Occurrence of AM Fungal Spores in Soils of *Tamarindus indica* L.

colonization also showed a great variation in all the localities during different seasons in different plant species. The similar observations made by, Gerdemann and Trappe (1974), and Ammani *et al.* (1994) also reported high level of AM colonization with low number of AM fungal spore population from various habitats.

The variation may be the result of variable host susceptibility, diverse type of AM fungi in the rhizospheric soils of individual plant species, host efficiency in soil resources capture of utilization (Koide, 1991; Clark and Zeto, 2000). On the other hand this may be due to soil type and quality (Ramann and Gopinathan, 1992) and other edapho–climatic factors (Abbott and Robson, 1991). According to the Bever (2002) found that AM fungal species, though associating with host cell have specific differences in their population and the growth rates. Abbott and Gazey (1994) reported that fungi differ in the manner and extent to which they colonize roots may due to root morphological nature and distribution of parenchyma cells in the cortex regions.

All the sixty four plants showed the presence of AM fungi, but the extent of colonization may vary. Throughout the study a root segment was considered as important organ for mycorrhizal association, even if any of three structures, *i.e.* hyphae, arbuscules and vesicles were present. Therefore, one study is not enough to generalize the AM fungi species composition of these plants. The AM fungi species comparison is known to change with time and crop (Schenck and Kinloch, 1980). Thus, extensive sampling over a long period is required to determine the species diversity. It needs to have better understanding to benefit the reflection and programme. These fungi should be evaluated for colonization and potential efficiency in singly and in combination

References

Abbott, L.K. and Robson A.D. (1991). Factors influencing in formation of AM. *Agriculture Ecosystem and Environment*, **35**: 121-150.

Abott, L.K. and Gazey, C. (1994). An ecological view of the formation of VA Mycorrhiza. *Plant and Soil*, **159**: 69-78.

Allen, E.B. and Allen, M.F. (1980). Natural re-establishment of VAM following strip mining reclamation in Wyoming *J. Appl. Ecol.*, **17**: 139-147.

Ammani, K., Venkateswarlu, K. and Rao, A.S. (1994). Vesicular arbuscular mycorrhizae in grasses: Their occurrence, indentity and development. *Phytopathology*, **6**: 397-418.

Bakshi, B.K. (1974). Mycorrhizal and its role in forestry project report PL 480. Dehradun: Forest Research Institute, 89 pp.

Bever, J.D. (2002). Host specificity of AM fungal population growth rates can generate feedback on plant growth. *Plant and Soil*, **244**: 281–390.

Byrareddy, M.S. and Bhagyaraj, D.J. (1988). Selection of effieicnt VAM fungal for inculating Leucaena in ixisols and vartisols. In: *Proceedings of the First Asian Conference on Mycorrhizal*, (Eds.) Mahadevan, N. Ramann and K. Natarajan. Madras University of Madras (29-31 January), Madras, India, pp. 271-273.

Chin K. Ong and Huxley, P. (1996). *Tree-Crop Interactions: A Physiological Approach.* CAB International, pp. 386.

Clark, R.B. and Zeto, S.K. (2000). Mineral acquisitation by arbuscular mycorrhizal plants. *Journal of Plant Nutrition*, **78**: 367-902.

Draf, M.J. and Nicolson, T.H. (1974). Arbuscular mycorrhizal fungi in colonizing coal waste in Scotland. *New Physol.*, **173**: 1129-1138.

Durga, V.V.K. and Gupta, S. (1995). Effect of vesicular-arbuscular mycorrhizae on the growth and mineral nutrition of Teak (*Tectona grandis* Linn. f.). *The Indian Forester*, **121**: 518-527.

Gerdemann, J.W. and Nicolson, T.H. (1963). Spore of mycorrhizal endogone species extracted from soil by wet sieving and decanting. *Trans. Br. Mycol. Soc.*, **46**: 235-244.

Gerdemann, J.W. and Trappe, J.M (1974). The endogonaceae of the Pacific Northwest, Mycological Memoir No.5. *Mycological Society America*, New York Botanical Garden, pp. 1-76.

Janson, D.P. (1980). Vesicular Arbuscular Mycorrhizae affected low land tropical rainforest plant growth. *Ecology*, **61**: 151-162.

Jasper, L.K., Abbott, L.K. and Robson, A.D. (1989). Academic response to additions of phosphorous of inoculation VAM fungi in soil stock pilled during mineral sand mining. *Plant and Soil*, **115**: 99-108.

Jorgonsen, M.J. (1979). Microorganisms of veclamation of mine waste. In: *Forest Soil and Landure*, (Ed.) C.T. Youngberg, pp. 251-268.

Koide, R.T. (1991). Nutrient supply, nutrient demand and plant response to mycorrhizal infection. *New Phythologist*, **11**: 35-44.

Kumar, P.P., Reddy, S.R. and Reddy, S.M. (2000). Mycorrhizal dependency of some agroforestry tree species. *New Phytologist*, **11**: 35-44.

Lakshman, H.C., Rajanna, L., Inchal, R.F., Mulla, F.I. and Srinivasulu, Y. (2001). Survey of VA mycorrhizae in agroforestry and its implication on forest trees. *Tropical Ecology*, **42(2)**: 283-286.

Lakshman, H.C. (2008). Arbuscular Mycorrhizal fungal diversity of tree species in dry deciduous forests of Dharwad. In: *Forest Biodiversity*, Vol. II. (Eds.) K. Muthuchelian, S. Kannaiyan and A. Gopalan. Associated Publishing Company, New Delhi, pp. 223-239.

Lakshman, H.C. (1996). VA Mycorrhizal fungi occurrence in some important timber tree species of kalagatagi forest in Karntaka. *J. Bihar. Bot. Soc.*, **5 (1-2)**: 16-22.

Lakshman, H.C. (1996). VA Mycorrhizae stadies in some economically important tree species. *Ph.D. Thesis*, Karnatak University, Dharwad.

Mukerji, K.G. (1996). Taxanomy of edomycorrhizal fungi. In: *Advance in Botany*, (Eds.) K.G. Mukerji, Binny Mathus, B.P. Chamold and P. Chitralekha. APH Publishing Corporation, New Delhi, pp. 212-218.

Muthukumar, S. and Udaiyan, K. (2000). Arbuscular mycorrhizal of plants growing in western ghats region South India. *Mycorrhiza*, **9**: 297-313.

Onguena, N.A. and Kayper, T.W. (2001) Mycorrhizal association in the rain forest of South Cameroon. *For Ecol. Mange*, **140**: 277-287.

Philips, J.M. and Hayman, D.S. (1970). Improved production for clearing roots and staining parasitic and vesicular arbuscular mycorrhizal fungi for rapid assessment of infection. *Trans Br. Mycol. Soc.*, **55**: 158-161.

Prashant, G. Inamdar and Lakshman, H.C. (2005).Symbiotic response of *Eugenia bracteata* Lam.To VA-Mycorrhial inoculation with different soil pH. *National Journal of Life Sciences*, **2(1&2)**: 161-164.

Rachel, E.K. Reddy, S.R. and Reddy, S.M. (1993). Effect of different mycorrhizal treatments on the growth and yield of sunflower. *Indian J. Microbial. Ecol.*, **3**: 99-103.

Raghupathy, S. and Mahadevan, A. (1993). Distribution of vesicular arbuscular mycorrhizal in plants and rhizosphere soils of tropical plains. Tamil Nadu, India, *Mycorrhiza*, **3**: 123-136.

Rahagdale, R. and Gupta, Nibha (1990). Vesicular arbuscular mycorrhizal association of biomass tree species in the tropical forest of Madhya Pradesh. *I. J. Forestry*, **22**: 62-65.

Ramana and Gopinathan, S. (1992). Association and activity of VAM of tropical trees in a tropical forest of Southern India. *Journal of Tropical Forest*, **8**: 311-312.

Reena, J. and Bagyaraj, D.J. (1999). Growth stimulation of *Tamarindus indica* by selected V-A-mycorrhizal fungi. *World. J. Microbiol. Biotechnol.*, **6**: 59-63.

Reeves, F.B., Wangner, D., Moorman, T. and Kiel, T. (1979). The role of endomycorrhizal in revegation practices in the semi and waste. IA comparison of incidence of mycorrhizal in several disturbed natural environments. *Amer. J. Bot.*, **66**: 6-13.

Schenk, N.C. and Perez, Y. (1990). *Manual for the Identification of VA Mycorrhial Fungi.* Synergistic Publication, USA.

Schenk, N.C. and Kinloch, R.A. (1980). Incidence of mycorrhizal fungi as six field crops in monoculture as newly created wood land site. *Mycologia*, **72**: 445-446.

Sharma, S.K., Sharma, G.D. and Mishra, R.R. (1986). Status of mycorrhizal in subtropical forest ecosystem of Meghalaya. *Acta Botanic India*, **14**: 87-92.

Stathi, P.D., Williams, S.E. and Christenson, M. (1988). Efficacy of native vesicular arbuscular mycorrhizal fungi after several soil disturbance. *New Phytol.*, **110**: 347–354.

Thapar, H.S. and Khan, S.M. (1973). Studies on endomycorriza in some forest species. *Proceedings of Indian Natural Science Academy* (*Part B; Biological Science*), **39(6)**: 687–694.

Young, A. (1997). *Agroforestry for Soil Management*. CAB International, pp. 320.

Zak, J.C., Danielson, R.M. and Parkinson, D. (1982). Mycorrhizal fungal spore number and species occurrence in two amended mine spoils in Alberta, Canda. *Mycology*, **74**: 347-354.

Chapter 17

Determining *Kunapajala* Pre-Sowing Treatment for Optimum Germination of Tomato (*Lycopersicon esculentum* L. cv. Selection 22) Seeds

***Rupali S. Deshmukh*[1*], *Neelam A. Patil*[1] *and T.D. Nikam*[2]**

ABSTRACT

Liquid fertilizer is a concept of modern agriculture for better yield. It is amazing to know that a better option for the liquid fertilizers is described in ancient Indian literature with more scientific formulation under the generic name '*Kunapajala*'. This is a fermentation product of easily available ingredients like *Sesamum indicum* L., bone marrow, flesh, black gram (*Vigna mungo*), milk, ghee

1 Post Graduate Research Centre in Botany, Tuljaram Chaturchand College, Baramati – 413 102 Dist. Pune, M.S., India

2 Department of Botany, University of Pune, Pune – 411 007, M.S., India

* Corresponding Author E-mail: drupali77@yahoo.com

and honey. However, there is no any documented reference on the influence of *kunapajala* on seed germination. In previous experimentation, 0.5 per cent *kunapajala* treatment was found to be useful for increasing plant growth. So, the present investigation is made to fix the period of pre-sowing treatment of *kunapajala* (0.5 per cent) for better germination of seeds of tomato (*Lycopersicon esculentum* L. cv. Selection22). Treatments are given at time intervals of 6 hours, 12 hours, 18 hours and 24 hours. Various parameters like germination per cent, root length, shoot length, vigour index, biomass, emergence index, speed of germination, coefficient of velocity of germination and number of secondary roots was studied. Overall picture shows that 18 hrs presoaking treatment of *kunapajala* (0.5 per cent) is useful to increase germination per cent, root length, shoot length, vigour index, biomass, emergence index, speed of germination and coefficient of velocity of germination. It is concluded that 18 hours treatment of *kunapajala* (0.5 per cent) gives the best results for the germination of tomato seeds.

Keywords: *Biomass production, Kunapajala, Lycopersicon esculentum.*

Introduction

The theme of liquid bio-fertilizers is mentioned in ancient Indian literature under the generic name "*Kunapajala*" by Kashyapa (800-900 AD) in 'Brihatsamhita', Surapala (1000 AD) in 'Vrikshayurveda' and Chakrapani Mishra (1577 AD) in 'Vishvavallabha'. *Kunapajala* is a fermentation product using easily available ingredients like *Sesamum indicum* L. (Tila), bone marrow, flesh (sheep, goat, fish etc), milk, black gram (*Vigna mungo*), ghee, honey etc. The beauty of *kunapajala* is that, it can be used on any plant at any growth stage. However, there is no fixed proportion for the ingredients of *kunapajala*. So, in previous experimentation (Deshmukh *et al.*, 2010) an attempt was made to standardize the procedure for preparation of *kunapajala*. It was found that 0.5 per cent treatment of *kunapajala* was useful for increasing tomato plant growth. Patil (2007) worked on Vrikshayurveda under the project entitled "Techniques in Ancient Indian Literature for Increasing Crop Productivity" which gave impetus to research on *kunapajala*. Mishra (2007) pointed out the need to standardize dose and time frequency for *kunapajala* administration.

The vegetables are important in human diet for their vitamin and mineral content essential for metabolic processes taking place within the human body. They confer disease resistance and make the body strong. Dieticians recommend inclusion of 300 gms of vegetables per day in human diet for balanced and nutritive diet, of which 115 gms should be fruit vegetables (Raul, 1999).Tomato is Solanaceous fruit vegetable available round the year. It is a source of carbohydrates, proteins, and fats. It contains vitamin A, vitamin B, vitamin C, salts, phosphorus, iron, potassium etc. along with malic acid which is useful to decrease cholesterol level. It is used to cure piles, skin diseases, chronic constipation; acidity etc.

Tomato seeds are very sensitive during germination especially during imbibitions of water. This is particularly seen when the sowing conditions are cold and wet. This sensitivity seems to be induced by the state of plasma membranes which regulate water penetration more or less efficiently (Fougereux, 2000). *Kunapajala* increases the vegetative and reproductive growth in tomato (Deshmukh *et al.*, 2010) however; there is no any documented reference on the influence of *kunapajala* on seed germination. With this back ground, an attempt was made in the present investigation to fix the period of treatment of *kunapajala* (0.5 per cent) for better germination of tomato (*Lycopersicon esculentum* L. cv. Selection 22).

Materials and Methods

The experiments were carried out at P.G. Research Centre, Department of Botany, Tuljaram Chaturachand College, Baramati, Dist. Pune, (M.S.) India, on seeds of tomato (*Lycopersicon esculentum* L. cv. Selection 22) obtained from 'Navalakha Seeds, Pune'. The seeds were surface sterilized with 0.05 per cent $Hgcl_2$ for 1 min., washed under running water and rinsed with distilled water. The seeds were allowed to germinate using Petri plate method. The seeds kept in distilled water for 6 hours worked as control. The experiments were carried out in triplicate.

Kunapajala was prepared as per the formulations given by Surpala (1000 AD) in Vrikshayuraveda verses 101 to 106 (Sadhale, 1996) and Chakrapani Mishra (1577 AD) in 'Vishvavallabha' verses 1, 2 and 3 respectively (Sadhale, 2004). Flesh of goat and pieces of fish (each 250 gm) were cooked in earthen pot with 1000 ml water. Then, 500 ml milk was added along with 500 gm each of seasum oil cake, ghee, cooked black gram and honey. It was stirred well and the fire was stopped. The lid was placed and the pot was placed for fermentation for 15 days in a warm place. The fermented liquid was filtered through cotton. Then the final volume was made 1000 ml (standard solution). It was stored in refrigerator. For pre-soaking treatment of seeds of tomato, 0.5 per cent solution was used (working solution). Treatments were given at time of intervals of 6 h, 12 h, 18 h and 24 h respectively.

The various parameters like germination percentage, length of radical, that of plumule and biomass were measured on 15 DAS using routine laboratory methods. Vigour Index (VI) was calculated according to the method suggested by Baki and Anderson (1973) Emergence Index (EI) was calculated by the following formula given by Baskin (1969). Speed of Germination (SG) was calculated by the formula given by Maguire (1962). Coeffient of Velocity of Germination (CVG) was calculated by the formula given by Kotowski (1962). Mobilization Efficiency (ME) of reserve food material present in seed during germination was calculated by method as described by Srivastava and Sareen (1974).

Results and Discussion

Seed priming is an easy technique to improve germination performance. Pre-soaking treatment by hydro-priming is actually osmo-conditioning, a physiological method which improves seed performance and provides faster and synchronized germination (Siveritepe and Dourado, 1995). Heydecker and Coolbaer (1997) mentioned that seed priming has been successfully demonstrated to improve germination and emergence in seed of vegetables. Nascimento (2005) remarked that primed seeds had higher germination compared to unprimed seeds especially at low temperature. In the present investigation seed priming with *kunapajala* gives the best results for the optimum germination of tomato seeds.

Table 17.1 shows the influence of pre-soaking treatments of 0.5 per cent *kunapajala* as on 15 DAS. In comparison with the control, all the four treatments (6 h, 12 h, 18 h and 24 h) were useful to enhance the seed germination. In both 12 h and 18 h treatments, increase in seeds germination, speed of germination and coefficient of velocity of germination was maximum (62 per cent each). But root length, shoot length and biomass showed maximum increase of 60 per cent, 44 per cent, and 28 per cent respectively with 18 h treatment. Mobilization efficiency and number of secondary roots showed maximum increase in 6 h treatment (156 per cent and 167 per cent) as against 12 h (112 per cent and 67 per cent), 18 h (113 per cent and 133 per cent) and in 24 h (24 per cent and 33 per cent) treatments of 0.5 per cent *kunapajala*. This is evident in Plate 17.1 which shows influence of pre-sowing treatment with *kunapajala* (6 h, 12 h, 18 h and 24 h) on germination of tomato seeds.

Vigour Index (VI) is the best criterion to assess the effect of any external agent on seed germination and seedling growth because it is calculated on the basis of germination percentage, root length and shoot length. Present investigation assures higher vigour index by presoaking treatment with *kunapajala*. The same is true for emergence index, speed of germination, coefficient of velocity of germination. Patel and Saxena (1994) mentioned that there are many reports which indicate increase in seed germination percentage and seedling vigour using plant growth regulators like gibberellic acid, kinetine, naphthalielic acid and indol butyric acid in green gram and black gram. Seed germination is highly orchestrated process. It is regulated by

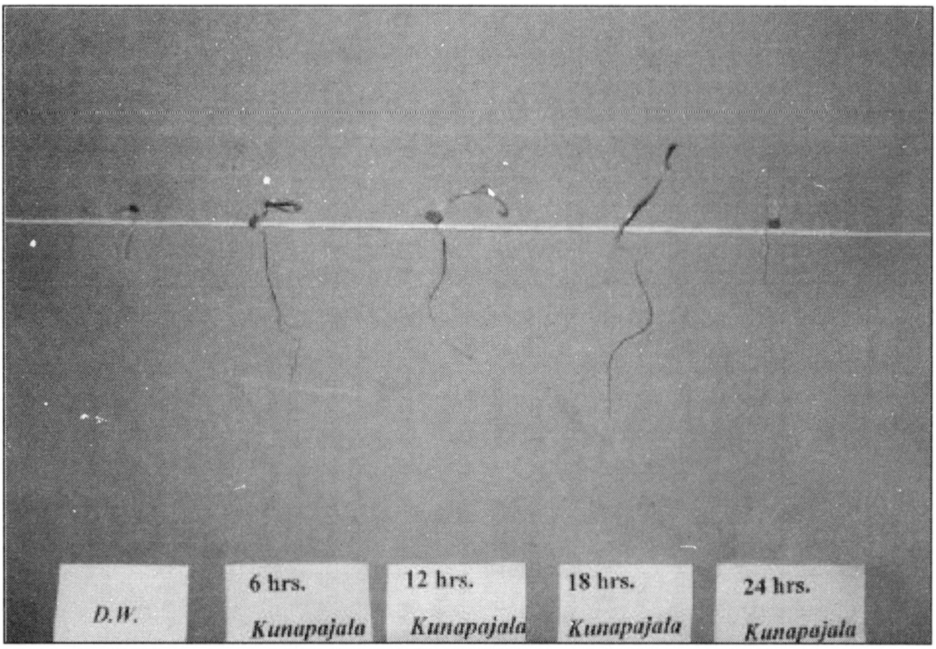

Plate 17.1: Influence of Pre-Soaking Treatment with *Kunapajala* (6 h, 12 h, 18 h and 24 h) on Germination of Tomato (*Lycopersicon esculentum* L. cv. Selection 22) Seeds

Table 17.1: Influence of Pre-Soaking Treatments of *Kunapajala* (6h, 12h, 18h and 24h) on Germination of Tomato (*Lycopersicon esculentum* L. cv. Selection 22) Seeds

Sl.No.	Parameters	Control	6h. Treatment of Kunapajala	% increase Over Control	12h. Treatment of Kunapajala	% increase Over Control	18h. Treatment of Kunapajala	% Oncrease Over Control	24h. Treatment of Kunapajala	% increase Over Control
1.	Germination%	15±0.5	30*±0.5	50	40*±0.5	62	40*±0.5	62	30*±0.5	50
2.	Root length (cm)	0.56±1.427	1.3*±2.397	56	1.435*±1.953	54	1.22*±2.003	60	0.3^ns^±1.921	-86
3.	Shoot length (cm)	0.72±1.893	1.27*±2.091	43	1.14*±1.614	36	1.3*±2.190	44	0.78*±1.540	8
4.	Vigour index	19.2±1.5	102.8**±1.55	435	103.4**±1.3	439	100.8**±1.2	425	32.4*±1.5	69
5.	Biomass (gm)	0.181±0.1	0.184*±0.01	2	0.226*±0.04	19	0.253*±0.07	28	0.186*±0.08	3
6.	Mobilization efficiency	15.6±0.1	40.0**±0.1	156	33.0**±0.1	112	33.3**±0.1	113	19.4*±0.1	24
7.	Emergence Index	3.448±0.6	4.16*±0.6	21	9.607**±0.6	179	10.101**±0.6	193	5.66*±0.6	64
8.	Speed of germination	0.2±0.09	0.4*±0.05	50	0.53*±0.01	62	0.53*±0.01	62	0.4*±0.04	50
9.	Coefficient of velocity of germination	0.066±0.5	0.133*±0.5	50	0.176*±0.5	62	0.176*±0.5	62	0.133*±0.5	50
10.	Number of secondary roots	0.3±2.394	0.8**±2.000	167	0.5*±1.427	67	0.7**±1.923	133	0.4*±1.542	33

The values are mean of twenty determinations.±: Standard deviation.

*: Significance at 5% level, **: Significance at 1% level, ns: Non-significance.

the interaction of plant growth regulators like GA, IAA and ABA (Taiz and Zeiger, 2002). Ruiz *et al.* (2006) found that, 0.05 per cent gibberellic acid induced the activity of endogenous hormones involved in seed germination and caused increase in seed germination of *Bromus auleticus*. The results of the present investigation indicated that *kunapajala* treatment might induce the activity of endogenous hormones involved in seed germination. Atiyeh *et al.* (2000) stated that tomato plants grown in vermicompost increased the rate of germination. An improvement in germination and growth of petunias, marigolds, bachelor buttons, poinsettias, bell pepper and tomatoes was observed in response to vermicompost (Subler *et al.*, 1998). The results of present investigation are also supported by observations of Thangaraj *et al.* (2009), who mentioned that organic liquid fertilizers enhanced the percentage and rate of germination in seeds of *Trigonella foenum-graecum*. Nene (2006) mentioned that, there is no fixed proportion for the ingredients of *kunapajala* and further research is needed to standardize the procedure and test it on crops. Neff *et al.* (2003) pointed out that, the effectiveness of *kunapajala* is due to the fact that the ingredients of *kunapajala* have been fermented, *i.e.* proteins, fats, carbohydrates etc. are broken into simple low molecular weight products. Therefore, nutrients from *kunapajala* become available to the plants faster than those from the traditionally applied organic matter. That is why *kunapajala* gives best results. The overall picture shows that among the various pre-soaking time periods of 0.5 per cent *kunapajala*, 18 hours treatment gives the best results for the germination of tomato seeds. The beauty of investigation is that the enhancement in germination parameter achieved by *kunapajala* treatment is in most natural way by supplying the nutrients to the plant which supports germination but does not alter plant metabolism.

Conclusion

From above study, it is concluded that the use of *kunapajala* (0.5 per cent) enhances the germination of tomato seeds and 18 hours pre-soaking treatment gives best results for better germination in seeds of tomato.

References

Atiyeh, R.M., Arancon, N.Q., Edwards, C. A. and Metzger, J.D. (2000). Influence of earthworm-processed pig manure on the growth and yield of greenhouse tomatoes. *Bioresource Technology*, **75(3)**: 175-180.

Baki, A. and Anderson, J.D. (1993).Vigour determination in soybean seed by multiple criteria. *Crop Science*, **3**: 630-663.

Baskin, C.C. (1969).GADA and seedling measurement as tests for seed quality. *Proceedings of Seedsmen, Mississippi State University*, 59-69.

Deshmukh, R.S., Patil, N.A. and Nikam, T. D. (2010). *Kunapajala*: Formulation and Standardization of Dosages for Cultivation of Tomato. *Proceedings of International Conference on Traditional Practices in Conservation Agriculture.* Udaipur, Rajasthan, India, pp. 125-132.

Fougeruex, Jean-Albert (2000). Germination quality and seed certification in grain legumes. *Grain Legumes*, **27(1)**: 14-16.

Heydecker, W. and Coolbaer, P. (1977). Seed treatments for improved performance survey and attempted prognosis. *Seed Science Technology*, **5**: 353-425.

Kotowski, F. (1962). Temperature relations to germination of vegetable seeds. *Proceedings of American Society of Horticultural Science*, **23**: 176-177

Maguire, J.D. (1962). Speed of germination aid in selection and evaluation for seedling emergence and vigour. *Crop Science*, **2**: 176-177.

Mishra, P. K. (2007). Effect of *Kunapa Jalam Vrikshyurveda* on growth of paddy. *Indian Journal of Traditional Knowledge*, **6(2)**: 307-310.

Nascimento, W.M. (2005). Vegetable seed priming to improve germination at low temperature. *Horticultura Brasileira*, **23**: 211-214

Neff, J. C., Chapin III F.S. and Vitousek, P. M. (2003). Breaks in the cycle: dissolved organic nitrogen in terrestrial ecosystems. *Frontiers in Ecology and the Environment*, **1(4)**: 205-211.

Nene, Y. L. (2006). Utilization of Traditional knowledge in agriculture. *Journal of Traditional Knowledge System of India and Sri Lanka*. Asian Agri History Foundation, Secunderabad, India: 32-38.

Patel, I. and Saxena, O.P. (1994). Screening of PGRs for seed treatment in green gram and black gram. *Indian Jr. of Plant Physiology*, **37(3)**: 206-208.

Patil, N.A. (2007). Techinques In Ancient Indian Literature for increasing crop productivity. Final Report of UGC Research Project.

Patil, N.A., Chiatale, R.D. and Dhumal, K.N. (2008). Role of oxygenated peptone in enhancing germination of tomato, brinjal and chilly. *Indian Journal of Plant Physiology*, **13**: 137-142.

Raul, V.G. (1999). Bhajipala Lagwad. Continental Prakashan, Pune, India: 42-56.

Ruiz, M. A., Perez, M. A. and Arguello, J.A. (2006). Conditions and stimulation for germination in *Bromus auleticus* seeds. Seed Sci. Technol. **34**. 19-24.

Sadhale, N. (Translator) (1996). Surapala's Vrikshayurveda (The Science of Plant Life). Asian Agri History Bulletin-1. Asian Agri History Foundation (AAHF), Secunderabad, Andhra Pradesh, India: 48-55.

Sadhale, N. (Translator) (2004). *Vishvavallabha* (*Dear to the world: the science of plant life*). Asian Agri History Bulletin-**5**. Asian Agri History Foundation (AAHF), Secunderabad, Andhra Pradesh, India: 76-80.

Siveritepe, H.O. and Dourado, A.M. (1995). The effect of priming treatment on the viability and accumulation of chromosomal damage in aged pea seeds. *Annals of Botany*, **75**: 165-171.

Srivastava, A.K. and Sareen, K. (1974). Physiology and biochemistry of deterioration of soybean seeds during storage. *Plant Horticulture*, **7**: 545-547.

Subler, S., Edwards, C.A. and Metzger, J.D. (1998). Comparing vermicomposts and composts. Bio Cycle. **39**: 63-66.

Taiz, L. and Zeiger, E. (2002). *Plant Physiology*. 3rd Edition, Panima Publishing Corporation, New Delhi.

Thangaraj, R., Sultan, A. I. and Srikumar (2009). Influence of Organic Fertilizer (Vermiwash, Effective Microorganisms and Panchagavaya) on the Germination of seeds of *Trigonella foenum-graecum* L. *Journal of Natural Science and Technology Life Science and Bioinformatics*, **1**: 108-113.

Chapter 18

Addition to the Flora of Marathwada Region of Maharashtra, India

S.P. Gaikwad[1], R.D. Gore[1] and K.U. Garad[1]*

ABSTRACT

Present paper deals with by reporting nine taxa of flowering plants from Ramling wildlife sanctuary and its adjoining region, as new records for Marathwada region of the Maharashtra State.

Keywords: Addition, Marathwada. Nine flowering plant taxa.

Introduction

The first exhaustive information about plant wealth of Marathwada region was furnished by Naik (1998). He has reported 1645 species and about 73 intraspecific taxa belonging to 746 genera and 155 families of the flowering plants from Marathwada region in his comprehensive book 'Flora of Marathwada'. In fact, Almeida (1996, 1998, 2001, 2009), Lakshminarasimhan (1996) and Singh and Karthikeyan (2000 and 2001) while dealing with their 'Flora of Maharashtra' have dealt with flowering plants of Marathwada region. Cooke (1958 Repr. ed.) in his

---

1 Life Science Research Laboratory, Walchand College of Arts and Science, Solapur – 6, Maharashtra, India

* Corresponding Author E-mail: sayajiraog@gmail.com

'Flora of Bombay Presidency' had not included Marathwada region, as it was under the Hyderabad State.

Ramling wildlife sanctuary and adjoining region holds an interesting area for floristic exploration because hills and hillocks of varying heights are scattered throughout the area, which support rich tropical dry deciduous and scrub vegetation. However, very little attention has been paid in past to assess the floristic components of the region (Naik 1966, 1967, 1969, 1970, 1979). Floristic explorations conducted in last two years yieided in collection of many interesting plants, which on critical study and comparison with authentic specimens deposited at BSI, Pune revealed that nine taxa new to the flora of Marathwada. They are presented here in alphabetical order genus wise with correct nomenclature. Family name is provided in parenthesis. A brief description, phonological data, GPS data, distribution, habitat, specimens examined and note if any are also given for each taxon.

Taxonomic Acocunt

Ammannia nagpurensis Mathew and Nair in Bull. Surv. India 31: 158, f.1 a-d. (1989) 1992; Almeida, Fl. Maha. 2: 282. 1989; Diwakar in Singh and Karthikeyan (ed.) Fl. Maha. 2: 31. 2001. (Plate 18.1) (LYTHRACEAE)

Annual herbs; stems *c* 30 cm, 4-winged. Leaves simple, opposite decussate, linear oblong. Flowers axillary, in pedunculate cymes. Calyx campanulate, *c* 1mm long, 4-lobed. Petals 4, caducous, *c* 1mm, pink. Stamens 4; style distinct. Capsules globose. Seeds many, ovoid, brown.

Fls. and Frts.: October–November

Distribution: Malumbra (N17° 56' 28.62" E76° 1' 48.72") in Osmanabad district.

Specimen examined: Starky point (Nagpur), *Mirashi* 252, (Blat.), *RDG*- 280 dated 21.11.2010

Note: So far this species was known from only Nagpur region of the Maharashtra State, so that present collection has extended the range of distribution of the species.

Anacardium occidentale L. Sp. Pl. 383, 1753; Hook. f. Fl. Brit. India 2: 20. 1876; Cooke, Fl. Pres. Bombay 1: 192. 1958 (Repr. ed.); Almeida, Fl. Maha. 1: 286. 1996; Prasanna in Singh and Karthikeyan (ed.) Fl. Maha. 2: 581. 2001. (ANACARDIACEAE)

A small tree. Leaves simple, 4-11 x 2.5-7 cm, obovate or elliptic, apex rounded, base cuneate. Panicles 15-25 cm long, terminal. Flowers polygamous, yellow, with pink strips. Stamens double than petals. Style single; ovary superior, glabrescent. Fruits a nut, reniform on juicy receptacles.

Fls. and Frts.: January–March.

Distribution: Apsinga (N18° 1' 40.53" E76° 3' 11.75") in Osmanabad districts

Specimen examined: Malvan, *A. N. Londhe* 191408, 15.12.2004 (BSI Pune), *RDG*- 60 dated 21.03.2010.

Note: It is cultivated as fruit tree.

Cuscuta campestris Yuncker in Mem. Torr. Bot. Club. 18: 138, f. 14. 1932; Almeida, Fl. Maha. 3B: 301. 2001; Venkanna and Kothari in Singh and Karthikeyan (ed.) Fl. Maha. 2: 490. 2001. (Plate 18.1) (CUSCUTACEAE)

Stem slender, yellowish. Flowers greenish yellow. Calyx lobes acute; pedicels shorter than flowers. Corolla lobes, acute; scales ovate, triangular. Stamens slender, infrastaminal scales present. Styles two; stigma globose.

Fls. and Frts.: November–February.

Distribution: Apsinga (N18° 4' 5.29" E76° 2' 7.63"), Bembli (N18° 9' 18.85" E76° 8' 27.34") in Osmanabad district.

Specimen examined: Near Patanwadi bridge, Pune, *Wadhwa* 64209, 12.09.1960; Koregaon park, Pune, *N. P. Singh* 112858 (BSI Pune), *RDG*- 351 dated 12.11.2010

Note: Occasionally parasite on *Achyranthus aspera* L., *Lantana camara* Linn. and *Vitex negundo* L. etc.

Eulophia graminea Lindl. Gen. Sp. Orchid. 182. 1833; Hook. f. Fl. Brit. India 6: 2. 1890; Bachulkar and Yadav in J. Econ. Tax. Bot. 17: 329. 1993. Lakshmi. in Sharma *et al.*, Fl. Maha. (Monocot.) 28. 1996; Almeida, Fl. Mah. 5A: 47. 2009. (Plate 18.1) (ORCHIDACEAE)

Pseudobulbs 5-10 x 5-9 cm, conical, pyriform to almost spherical, green, marked with transverse lines of leaf bases. Leaves linear with sheathing leaf bases, lower reduced to sheaths, upper 7-27 x 1-5-2.5 cm. Scapes lateral, 1-2 per pseudobulb, 50-90 x 0.5-0.7 cm. Flowers in racemes. Capsules 2-2.5 x 0.7-0.8 cm, ellipsoid-oblong, drooping.

Fls. and Frts.: September–April.

Distribution: Apsinga (N18° 5' 0.33" E76° 1' 40.44") in Osmanabad district.

Specimen examined: *RDG*- 78 dated 02.04.2010

Note: Bachulkar and Yadav (1993) were reported this orchid from sugarcane fields near Islampur, dist. Sangli but they had seen only two individuals. In light of this present collection confirms occurrence this orchid in Maharashtra State. However, it was found uncommon on wet margins of temporary running water streams.

Lavandula bipinnata O. Ktze. Rev. Gen. Pl. 521.1891; Kulkarni and Das Das in Singh and Karthikeyan (ed.) Fl. Maharashtra 2: 722. 2001. *Lavendula burmanni* Bth. Lab. Gen. and Sp. 151. 1833; Hook. f. Fl. Brit. India 4: 631. 1885; Cooke, Fl. Pres. Bombay 2: 534. 1958 (Repr. ed.). (LAMIACEAE)

var. *bipinnata*

Erect herbs, 20-25 cm tall. Stem slender, simple or branched, pubescent. Leaves 2-8 cm long, pinnatipartite, deeply lobed; lobes linear. Flowers bracteolate; bracts equal to or shorter than calyx, awned. Calyx tubular, hairy. Corolla Pale blue or white, 5 lobed, pubescent outside. Nutlets smooth, ellipsoid, Black.

Fls. and Frts.: October–February.

Plate 18.1

Eulophia graminea **Lindl.**

Lens culinaris **Medik.**

Physalis pubescens **Linn.**

Ammannia nagpurensis **Mathew and Nair**

Tragia involucrata **L.**

Cuscuta campestris **Yuncker**

Distribution: Ghatangri (N18° 13' 17.71" E76° 1' 15.26"), Apsinga (N18° 3' 17.28" E76° 3' 54.30") in Osmanabad district.

Specimen examined: *RDG*- 396 dated 05.12.2010

Lens culinaris Medik., Vorles. Churpf. Phys. Ocon. Ges. 2: 361. 1787; Sanj. Legumes of India 202. 1991; Almeida, Fl. Maha. 2: 102. 1998; Kothari in Singh and Karthikeyan (ed.) Fl. Maha. 1: 771. 2000. *Lens esculenta* Moench. Meth. Pl. 131.1794; Cooke, Fl. Pres. Bombay 1: 435. 1958 (Repr. ed.). (Plate 18.1) (FABACEAE)

Pubescent, sub-erect herb. Leaflets 4-6 paired, lanceolate, rachis terminating in to tendril, villous. Raceme 1–3 flowered. Flowers white or blue-purple. Pods compressed, 10–15 mm, smooth, yellow. Seeds 2, lenticular.

Fls. and Frts.: December–March.

Distribution: Bembli in Osmanabad district (N18° 9' 11.49" E76° 11' 53.79").

Specimen examined: Near Vaval dam, Pune dist. *B. V. Reddi* 97922, 26.06.64 (BSI Pune), *RDG*- 461 dated 02.03.2011

Note: It was found occasionally in sugar cane fields.

Neolamarckia cadamba (Roxb.) Bosser. Bull. Mus. Nat. Hist. (Paris) Sect. B, Adansonia 3: 247. 1984; Almeida, Fl. Maha. 3A: 41. 2001; Mudaliar and Prasad in Singh and Karthikeyan (ed.) Fl. Maha. 2: 155. 2001. *A. cadamba* (Roxb.) Miq. Fl. Ind. Bat. 2: 135. 1856; Hook. f. Fl. Brit. India 3: 23. 1880. *A. indicus* A. Rich. in Mem. Soc. Hist. Nat. Paris 5: 238. 1834; Cooke, Fl. Pres. Bombay 2: 6. 1958 (Repr. ed.) (RUBIACEAE)

A large tree. Leaves elliptic-lanceolate, coriaceous, acute or obtuse. Flowers in dense globose head, ebracteolate. Calyx tube not fused in to fleshy mass. Corolla lobes imbricate; ovary 4 celled in upper portion.

Fls. and Frts.: November–February

Distribution: Osmanabad town (N18° 11' 28.08" E76° 1' 46.78").

Specimen examined: Danoli (Ratnagiri dist.) *John Chevian* 105446, 26.06.1965; *R. K. Bhide s.n.* 22.09.1902 (BSI Pune), *RDG*- 334 dated 28.10.2010

Note: It is often cultivated around temples as religious tree.

Physalis pubescens Linn. Sp. Pl. 183. 1753; Almeida, Fl. Maha. 3B: 371. f. 234. 2001. *Physalis turbinata* Medic. in Act. Acad. Theod. Palat. 4(Phys.): 184, t. 4. 1780. (Plate 18.1) (SOLANACEAE)

Branched herbs. Stems woody, rough with white soft hairs all over. Leaves simple, broadly ovate, long petioled, margins narrowly serrate, hairy at both sides. Flowers solitary, axillary, on short thick pedicel, hairy. Calyx lobes with acuminate apex; calyx lobes exceeding the berry, calyx tube rounded. Fruit a berry. Seeds many, light orange, smooth, broadly heart shaped.

Fls. and Frts.: August–January.

Distribution: Pohaner (N 18° 6' 58.32" E76° 1' 6.53"), Osmanabad (N18° 11' 12.16" E76° 2' 17.52"), Apsinga (N18° 3' 40.73" E76° 2' 25.17") in Osmanabad district.

Specimen examined: Nasik road *H. Santapau 18378-9* (Blat.), *RDG*- 322 dated 24. 10.2010

Note: It is common along road sides and in waste land. The species is distinct by having stout stem and pubescence all over body.

Tragia involucrata L. Sp. Pl. 980. 1753; Hook. f. Fl. Brit. India 5: 465. 1888; Cooke, Fl. Pres. Bombay 3: 119. 1958 (Repr. ed.); Londhe in Singh and Karthikeyan (ed.) Fl. Maharashtra 2: 904. 2001. (Plate 18.1). (EUPHORBIACEAE)

Perennial herbs, hispid with stinging hairs; stem twining. Leaves oblong-lanceolate to broadly ovate, apex acuminate, base rounded or cordate, margins serrate hairy. Flowers short pedicillate in hairy racemes; male in upper part and Female in lower part of racemes. Capsule 3-lobed, white, more or less hispid. Seeds smooth, globose.

Fls. and Frts.: October–January.

Distribution: Naldurg in Osmanabad district (N17° 48' 23.76" E76° 17' 32.22").

Specimen examined: Ameni island, *Wadhwa* 49030, 25.02.1959 (BSI Pune), *RDG*-431 dated 02.01.2011

Acknowledgement

Authors are thankful to the Principal, Walchand College of Arts and Science, Solapur for providing available research facilities, Director, Botanical Survey of India, Western Circle, Pune for confirmation of identifications and to RGSTC, Govt. of Maharashtra for financial assistance.

References

Almeida, M.R. (1996). *Flora of Maharashtra.* Blatter herbarium, St. Xavier's College, Mumbai. **1**: 286.

Almeida, M.R. (1998). *Flora of Maharashtra.* Blatter herbarium, St. Xavier's College, Mumbai, **2**: 102 and 282.

Almeida, M.R. (2001). *Flora of Maharashtra.* Blatter herbarium, St. Xavier's College, Mumbai. **3a**: 41.

Almeida, M. . (2001). *Flora of Maharashtra.* Blatter herbarium, St. Xavier's College, Mumbai. **3b**: 301 and 371.

Almeida, M.R. (2009). *Flora of Maharashtra.* Blatter herbarium, St. Xavier's College, Mumbai. **5a**: 47.

Bachulkar, M.P. and Yadav, S.R. (1982). Some new plant records for Maharashtra. *J. Econ. Tax. Bot.*, **17**: 329.

Cooke, T. (1958). *Flora of Bombay Presidency* (Repr. ed.) Govt. of India (Bot. Surv. India). Vol. **1-3.**

Lakshminarasimhan, P. (1996). *Flora of Maharashtra, Monocotyledons,* series 2. in B.D. Sharma, S. Karthikeyan and N.P. Sharma (Eds.).*Bot. Surv. India.*, p. 28.

Naik, V. N. (1970). A census of *Crotalaria* species in Osmanabad district. *Marathwada Univ. J. Sci.* 9: 15-18.

Naik, V. N. (1966). A new *Crotalaria* species from Osmanabad district. *Ind. For.* **92 (12):** 790-791.

Naik, V. N. (1967). *Amaranthus polygonoides* L. from Osmanabad district, a new record for India. *J. Bombay. Nat. Hist. Soc.* **64(1):** 134-135.

Naik, V. N. (1969). An artificial key to the Leguminosae of Osmanabad district. *Marathwada Univ. J. Sci.* **8(1):** 15-19.

Naik, V. N. (1979). *Flora of Osmanabad*. Venus Publishers, Aurangabad.

Naik, V. N. (1998). *Flora of Marathwada*. Amrut Prakashan, Aurangabad, Vols. 1&2.

Sanjappa, M. (1991). *Legumes of India*. Bishen Singh and Mahendra Pal Singh, Dehradun, India., p. 202.

Singh, N. P and S. Karthikeyan (2000). *Flora of Maharashtra* Vol. I. series 2. Bot. Surv. India.

Singh, N. P and S. Karthikeyan (2001). *Flora of Maharashtra*. Vol. II. series 2. Bot. Surv. India.

Chapter 19

Diversity of Arbuscular Mycorrhizal (AM) Fungi in Some Common Plants of Marathwada

P.P. Sarwade[1]*, V.S. Sawant[1] and U.N. Bhale[1]

ABSTRACT

Four angiospermic plants belonging to two different families were studied for their AM association. All test plants were growing and distributed in Osmanabad district of Marathwada region in Maharashtra state. The result showed that all the four plants *viz. Annona squamosa, Annona reticulata, Tinospora cordifolia* and *Cocculus hirsutus* had AM fungal association in the roots and spore population in the rhizosphere soil. However, maximum percent root colonization of AM fungi was observed in *Tinospora cordifolia* (92 per cent) followed by others, while minimum in *Annona reticulata* (57.33 per cent). *Tinospora cordifolia* (320) showed more spore density whereas less in *Annona squamosa* (59). Total five genera of AMF was identified up to species level in which *Acaulospora* spp were found dominate followed by *Glomus* spp, *Sclerocystis* spp, *Entrophospora* spp. and *Gigaspora* spp were found poorly distributed.

Keywords: *AM fungi, Plants, Root colonization.*

---

1 Research Laboratory, Department of Botany, Arts, Science and Commerce College, Naldurg Tal. Tuljapur Dist. Osmanabad – 413 601, M.S., India

* Corresponding Author E-mail: ppsarwade@gmail.com

Introduction

Arbuscular mycorrhizal fungi (AMF) form a symbiotic association with majority of land plants improving plant growth. More than 80 per cent of all plants are associated with AMF in their root system (Smith and Read, 1997). These well established AMF contribute to the phosphorus nutrition of plants by enhancing phosphorus uptake from the soil (Draft and Nicolson, 1966). Marathwada region forms the part of the Vast Deccan Plateau of India and is one of the four divisions of Maharashtra state. The rivers and rivulets which are dry for major part of the year. This enormously powerful during monsoon and flow with great speed. This results loss of soil cover and exposes bare rocks at many places. Recently, their have been some serious efforts to control this enormous soil erosion. Until AMF is too low to contribute to the successful establishment of plant species which can help to improve recovery rate of the soil system.

Annona squamosa, Annona reticulata, Tinospora cordifolia and *Cocculus hirsutus* are multipurpose plant species commonly found in Maharashtra state. Hence a study survey was conducted around Osmanabad district in Marathwada region, where the plant is grown throughout the year to observe AM fungal genera and species that are associated with four plants.

Materials and Methods

Roots and rhizosphere soil samples of four plants (*viz. Annona squamosa, Annona reticulata, Tinospora cordifolia* and *Cocculus hirsutus*) were collected and in each plant three replications were taken. Root samples were brought to the laboratory which were then washed in tap water and cut in to 1 cm pieces in length. Root samples were cleared and stained using Phillips and Hayman (1970) technique. Root colonization was measured according to the Giovannetti and Mosse (1980) method. Hundred grams of rhizosphere soil samples were analyzed for their spore isolation by wet sieving and decanting method (Gerdmann and Nicolson, 1963). Identification of AM fungal genera up to species level by using the Manual for identification Schenck and Perez (1990).

Results and Discussion

The data of percent of colonization and spore number associated with four different plants are presented in Table 19.1.

The result shows that all the tested plants were colonized by AMF. Maximum percent of colonization were found in *T. cordifolia* (92 per cent) than other three plants whereas, minimum percentage was found in *A. reticulata* (57.33 per cent). Hyphal and vesicular types of colonization were found in roots of different plants. Hyphae were almost common in all tested plants. More number of spores (320) was observed in rhizosphere soil of *T. cordifolia* than *A. squamosa, A. reticulata* and *C. hirsutus*. Total five genera were observed *viz., Acaulospora* spp *Glomus* spp, *Sclerocystis* spp, *Entrophospora* spp and *Gigaspora* spp. Highest number of AMF genera and species was associated with *T. cordifolia* while the lowest number of AM fungal genera and species were recorded in other three plant species. Among AM fungal species

Acaulospora spp were found dominate followed by *Glomus* spp., *Sclerocystis* spp., *Entrophosphora* spp and *Gigaspora* spp were found poorly distributed.

Table 19.1: Per cent Root Colonization and Spore Population in Four Common Plants

Sl.No.	Plant Species	Family	*Colonization (%)	Types of Colonization	*Spore Population	AM Fungal Genera
1.	*Annona squamosa* L.	Annnona-ceae	63	H	59	*Glomus* spp. *Acaulospora* spp.
2.	*Annona reticulata* L.	Annnona-ceae	57.33	HV	217	*Glomus* spp. *Acaulospora* spp. *Gigaspora* spp.
3.	*Tinospora cordifolia* (Willd) Miers	Menisper-maceae	92	HV	320	*Glomus* spp. *Acaulospora* spp. *Sclerocystis* spp. *Entrophosphora* spp.
4.	*Cocculus hirsutus* (L.)	Menisper-maceae	62	HV	150	*Glomus* spp. *Entrophosphora* spp. *Acaulospora* spp.

* Mean of three samples; H: Hyphae; V: Vesicular.

Most plant species are typically mycorrhizal with approximately 4/5 of all land plants forming AM associated (Molloch *et al.*, 1980), because of greater efficiency in nutrient uptake from soil (Draft and Nicolson 1966). The occurrence of AMF in medicinal plants has reported earlier by Taber and Trappe (1982), Udea *et al.* (1992), Muthukumar and Udaiyan (2001), Selvaraj *et al.* (2001) and Rani and Bhaduria (2001). Recently, Bukhari *et al.* (2003), Muthukumar *et al.* (2006) and Swapana and Ammani (2009) reported the occurrence of AMF in medicinal plants from India.

The result obtained from the study suggests that the colonization percentage and number of AM fungal spores differ with different four common plants. Among the five genara *Acaulospora* spp was found much more frequent than other genera.. The highest number of mycorrhizal spores in rhizosphere soil and AM fungal infection in the roots of *T. cordifolia* indicated that these plant species might be considered good host for AMF under natural conditions. Therefore, here concluded that, occurrence or distribution of AMF varies with host ranges.

Acknowledgements

Authors are grately thankful to Principal ASC College, Naldurg and Prof B. F. Rodrigues, Goa University, Goa for their constant encouragement and valuable guidance.

References

Bukhari, M.J., Khade, S.W., Jaiswal, V.J., Gaonkar, U.C. and Rodrigues, B.F. (2003). Arbuscular mycorrhizal (AM) status of Tropical Medicinal plants: A field survey of Arbuscular mycorrhizal fungal association in Herbs. *Plant Archives*, **3 (2)**: 167-174.

Draft, M.J. and Nicolson, T.H. (1966). The effect of endogone mycorrhizae on plant growth. *New Phytologist*, **65**: 343-350.

Muthukumar, T. and Udaiyan, K. (2001). Vesicular arbuscular mycorrhizal association in medicinal plants of Maruthamalai Hills, Western Ghats, Southern India. *J. Mycol. Pl. Pathol.* **31(2)**: 180-184.

Muthukumar, T., Senthilkumar, M. and Rajangam, M. (2006).Arbuscular mycorrhizal morphology and dark septate fungal associations in medicinal and aromatic plants of Western Ghats, Southern India. *Mycorrhiza*, **17**: 11-24.

Malloch, D.W., Pirozynski, K.A. and Raven, P.H. (1980). Ecological and evolutionary significance of mycorrhizal symbiosis in vesicular plants. *Proceedings of National Science Academic*, **75**: 2113-2118.

Dwivedi, O.P. and Vyas, D. (2002). Effect of Potassium on the occurrence of AM fungi. *J. Basic and Appl. Mycol.*, **1**: 233-235.

Giovannetti, M. and Mosse, B. (1980). An evaluation of techniques of measuring vesicular arbuscular mycorrhizal infection in roots. *New Phytol.*, **84**: 489-500.

Gerdemann, J.W. and Nicolson, T.H. (1963). Spores of mycorrhizal *endogone* species extracted from soil by wet sieving and decanting. *Trans. Br. Mycol. Soc.*, **46**: 235-244.

Phillips, J.M. and Hayman, D.S. (1970). Improved procedures for clearing root and staining parasitic and vesicular arbuscular mycorrhizal fungi for rapid assessment of infection. *Tans. Bri. Mycol. Soc.*, **55(1)**: 158-161.

Rani, V. and Bhaduria, S. (2001). Vesicular arbuscular mycorrhizal association with some medicinal plants growing on alkaline soil of Manipuri District, Uttar Pradesh. *Mycorrhiza News.*, **13(2)**: 12-14.

Schenck, N. C. and Perez, Y. (1990). *Manual for the Identification of Vesicular Arbuscular Mycorrhizal Fungi*. Synergistic Publications: Gainesville, FL., U.S.A., pp. 1-286.

Smith, S.E. and Read, D.J. (1997). *Mycorrhizal Symbiosis*, 2nd edn. Academic, San Diego, CA.

Selvoraj, T.R., Murugan and Bhaskaran, C. (2001). Arbuscular mycorrhizal association of kashini (*Cichorium intybus* L.) in relation to physicochemical characters. *Mycorrhiza News.*, **13 (2)**: 14-16.

Swapna, V. L. and Ammani, K. (2009). Association of AM fungi on medicinal plants at coal mines of Mancheriala, Andra Pradesh. *Mycorrhiza News.* **21(2)**: 5-6.

Taber, T.A. and Trappe, J.M. (1982). Vesicular arbuscular mycorrhiza in rhizomes, scale like leaves, roots and xylem of ginger. *Mycologia.*, **74**: 156-161.

Udea, T., Husope, T., Kubo, S. and Nakawashi, I. (1992). Vesicular arbuscular mycorrhizal fungi (Glomales) in Japan II. A field survey of vesicular arbuscular mycorrhizal association with medicinal plants in Japan. *Trans. Br. Mycol. Soc.* **33**: 77-86.

Chapter 20

Preparation of Chloroplastic and Cytoplasmic LPC from Various Plant Species

V.G. Manwatkar[1]* *and D.P. Gogle*[1]

ABSTRACT

The leafy vegetables are consumed as a source of protein, vitamins and minerals. The green chloroplastic leaf protein concentrates (LPC) is rich in proteins, lipids, chlorophyll and xanthophylls pigments while the proteins associated with cytoplasmic LPC are nutritionally superior. Leaf protein concentrate (LPC) was extracted from green leaves of some wild and cultivated plants to prepare the two types of LPC samples from nine different plant species *viz. Brassica juncea* (L.) Czern. and Coss, *Brassica napus* L., *Brassica oleracea* var. *Botrytis* L. (cultivated) and *Chenopodium album* L, *Goniocaulon indicum,* (Klein ex Willd). C.B. CL, *Celosia argentea* L., *Vigna trilobata* (L.)Verde, *Digera muricata* (L.) Mart, and *Tridax procumbens* L (wild) by heat coagulation method. The highest amount of chloroplastic and cytoplasmic protein was found in *Brassica oleracea i.e.* 8.5g and 0.985g per 100 ml juice. *Brassica napus* and *Chenopodium album* showed 8.0g and 0.5g chloroplastic and cytoplasmic LPC. The lower amount of LPCs was observed in *Goniocaulon indicum i.e.* 4.0g and 0.49g. The remaining plant species exhibits 4.5g to 7.0g and 0.35g to 0.56g chloroplastic and cytoplastic LPC per 100ml of juice. The LPC extracted from these plant species was used for nutritional studies.

Keywords: *Cytoplasmic protein, Chloroplastic protein, Leaf Protein Concentrates (LPC).*

---

1 Department of Botany, RTM Nagpur University Campus, Nagpur – 440 033, M.S., India

* Corresponding Author E-mail: dilim_vijju11@yahoo.co.in

Introduction

The leafy vegetables are consumed as a source of protein, vitamins and minerals. Several workers have investigated to utilized leaves as a protein source in the past and developed methods of isolation, fractionation and utilization of known and underutilized sources of different green leaves from various species (Pirie, 1942; Morrision and Pirie, 1961; Akeson and Stahmann, 1965; Fernandez *et al.*, 2007). The green chloroplastic leaf protein concentrates (LPC) is rich in proteins, lipids, chlorophyll and xanthophyll pigments, while the proteins associated with cytoplasmic LPC are nutritionally superior.

Materials and Methods

In the present investigation attempts were made to prepare the two types of LPC samples from green leaves of nine different plant species selected from wild and cultivated which are abundantly available in the region. These plant species are *Brassica juncea* (L.) Czern. and Coss, *Brassica napus* L., *Brassica oleracea* var. *Botrytis* L. (cultivated) and *Chenopodium album* L, *Goniocaulon indicum*, (Klein ex Willd). C.B. CL, *Celosia argentea* L., *Vigna trilobata* (L.)Verde, *Digera muricata* (L.) Mart, *Tridax procumbens* L.

Pirie (1942) suggested the method for preparation of leaf protein concentrates (LPC). In this method, the plant leaves pulped and pressed and the expressed juice was heated to 60°C and subsequently at 95°C. After filtration, due to the heating at 60°C protein associated with chloroplast coagulate along with chlorophyll pigments resulting in to dark green LPC referred to as 'Chloroplastic LPC'. When green chloroplastic LPC is isolated from the juice and the filtrate remaining behind is heated at 95°C, the remaining protein from cytoplasm precipitate and coagulate, resulting in to yellowish to white protein concentrate referred to as 'Cytoplasmic LPC'.

Results and Discussion

The yield of various LPC preparations per unit volume of juice extracted from different plant species is given in Table 20.1. Significant differences in yield of different leaf protein concentrates like whole unfractionated, chloroplastic and cytoplasmic LPC's were observed and analysed on the basis of fresh and dry wt. (Per 100ml juice). The *B. oleracea* yielded higher amount of unfractionated LPC (9.485gm and 4.512gm), chloroplastic LPC (8.0gm and 4.0gm) and cytoplasmic LPC (0.985gm and 0.512gm) on fresh and dry wt. basis respectively. The species of *B. napus* and *C. album* also yielded more than 8.0gm (Fresh wt. basis) unfractionated and chloroplastic LPC. The cytoplasmic LPC was also found more than 0.5gm on fresh wt. and 0.2gm on dry wt. basis respectively.

The species like *B. juncea, G. indicum, C. argentea, V. trilobata, D. muricata* and *T. procumbens* yielded 7.356gm to 4.490gm unfractionated LPC on fresh wt. basis whereas 3.961gm to 1.376gm on dry wt. basis. The chloroplastic LPC was found in the ranged of 7.0gm to 4.0gm on fresh and 3.5gm to 1.0gm on dry wt. basis in which lower amount was observed in *G. indicum.* The yield of cytoplasmic LPC was found in the ranged of 0.782gm to 0.356gm on fresh wt. while 0.468gm to 0.172gm on dry wt.

Table 20.1: Preparation of Cytoplasmic and Chloroplastic LPC from Various Plants

Name of Plant	*Yield of LPC (gm/100ml Juice)*							
	Unfractionated LPC		*Chloroplastic LPC*		*Cytoplasmic LPC*		*Chloroplastic : Cytoplasmic Ratio*	
	Fresh Wt.	*Dry Wt.*	*Fresh Wt.*	*Dry Wt.*	*Fresh Wt.*	*Dry Wt.*	*Fresh Wt.*	*Dry Wt.*
Brassica juncea (L.) Czern. and Coss.	7.356	2.172	7.0	2.0	0.356	0.172	19.66:1	11.62:1
Brassica napus L.	8.541	3.327	8.0	3.0	0.541	0.327	14.78:1	09.17:1
Chenopodium album L.	8.533	4.217	8.0	4.0	0.533	0.217	18.43:1	15.00:1
Goniocaulon indicum (Klein ex willd). C.B. C.L.	4.49	1.376	4.0	1.0	0.49	0.376	08.16:1	02.65:1
Brassica oleracea var. *Botrytis* L.	9.485	4.512	8.5	4.0	0.985	0.512	08.62:1	07.81:1
Celosia argentea L.	6.579	2.838	6.0	2.5	0.579	0.338	10.36:1	07.39:1
Vigna trilobata (L.) Verde.	7.782	3.468	7.0	3.0	0.782	0.468	08.95:1	06.41:1
Digera muricata (L.) Mart.	5.731	3.961	5.0	3.5	0.731	0.461	07.59:1	06.83:1
Tridax procumbens L.	5.062	2.301	4.5	2.0	0.562	0.301	08.00:1	06.64:1
Mean	7.06	3.13	6.44	2.78	0.62	0.35		
Standard Deviation (S.D)	1.71	1.04	1.65	1.00	0.19	0.11		
Standard Error (S.E)	0.57	0.35	0.55	0.33	0.06	0.04		

basis. The minimum amount of cytoplasmic protein was found in *B. juncea*. The chloroplastic and cytoplamic ratio was found highest in *B. juncea* on fresh wt. basis whereas it was found higher in *C. album* on dry wt. basis. The higher value for chloroplastic LPC indicated high proportion of proteins associated with chloroplast and hence higher photosynthetic efficiency of these plant species. The result indicated that the yield of chloroplastic LPC was 7 to 19 times higher on fresh wt. basis and 2 to 15 times higher on dry wt. basis than cytoplasmic LPC.

In conclusion, *C. album* which is wild plant and Brassica species which are oil yielding crops can be successfully used for the extraction of chloroplastic and cytoplasmic LPC.

References

Akeson, W.R. and Stahmann, M.A. (1966). Leaf protein concentrates: A comparison of protein production per acre of forage with that from seed and animal crops. *Econ. Bot.* **20**: 244-250.

Fernandez, S.S., Menendez, C., Mucciarelli, S. and Padilla, A.P. (2007). Saltbush (*Atriplex lampa*) leaf protein concentrate by ultrafiltration for use in balanced animal feed formulations. *J. Sci. Food Agr.*, **87**: 1850-1857.

Morrison, J.E. and Pirie, N.W. (1961). The large scale production of protein from leaf extracts. *J. Sci. Food Agric.*, **12**: Pp.1- 5.

Pirie, N.W. (1942). Direct use of leaf protein in human nutrition. *Chem. Ind.*, **61**: 45.

Chapter 21

Effect of Different Concentrations of Chitosan on the Growth of Okra [*Abelmoschus esculentus* (L.) Moench.]

***Sandhya Jaybhay*[1], *Lata Done*[2] *and Avinash B. Ade*[3]**

ABSTRACT

In the present investigation different concentrations of chitosan i. e. 1ml (0.03 per cent) and 20ml (0.6 per cent) were used for seed treatment and the seeds were sown in pots in replicates with control. The pots were kept in sunlight. The growth was measured in terms of plant height and number of leaves, leaf area index and yield components. In low concentrations of chitosan (1ml) these quantities were found higher in the plants as compare to control.

Keywords: Chitosan, Seed treatment etc.

---

1 Department of Botany, Dr. Babasaheb Ambedkar Marathwada University, Aurangabad – 431 004, M.S., India

2 Department of Botany Arts, Science and Commerce College, Naldurg Tq.Tuljapur, Dist. Osmanabad – 431 602, M.S., India

3 Department of Botany, University of Pune, Pune – 411 007, M.S., India

Introduction

Okra *Abelmoschus esculentus* (L.) Moench is originated in south Asia. It is a vegetable crop. It contains mucilage and fiber. It is popular in India and Pakistan. It is medicinally used because of its diuretic properties (Felter *et al.*, 2007). Okra improves the overall health of digestive system. Okra contains some important nutritional compounds for human health. Okra mucilage could prove to be cost effective and efficient biodegradable alternative flocculants for textile effluent treatment (Srinivasan *et al.*, 2008). Chitosan is a natural biodegradable polysaccharide polymer is component of exoskeleton of crust ace and insects. It also found in the cell walls of variety of fungi. Chitosan control the diseases of rose (Wojdyla *et al.*, 2001). Spraying of chitosan significantly increased all vegetative parameters such as fresh and dry weight, photosynthetic pigments, yield and decreased disease (Tantwy *et al.*, 2009). Chitosan increased the photosynthetic pigment of maize and soybean (Khan *et al.*, 2002) chitosan increased the seed germination in maize (Shao *et al.*, 2005). In the present study the attempts have been made to study the effect of chitosan on okra with respect to morphological parameters.

Material and Methods

Okra seeds of variety, *Arka* and *Anamika* were collected from local market of Aurangabad (Maharashtra State). Chitosan solution of 0.03 per cent (1 ml) and 0.6 per cent (20 ml) were used for in each solution. Twenty seeds were soaked in 100 ml solution for 2 hours. After treatment these seeds were washed thoroughly with water and the seeds were sown in pots with replicates. The observations were recorded randomly by selecting 3 plants from each pot with a particular concentration. The height of the plants, number of leaves, leaf area and the 4th leaf separated and the plants were dried in hot air oven at 80 °C first one hour and then at 60 °C till the sample were completely dried. The biomass of the leaves was recorded. The number of flowers and number of fruits were recorded. ANOVA and the values for critical difference (C.D.) were calculated at $p = 0.05$.

Results and Discussion

It is cleared from Table 21.1 that very low concentration *i.e.* 1ml (0.03 per cent) chitosan treatment increased the height of the plant and slightly increased the no of leaves as compare to high concentration of chitosan and control. It is cleared from Table 21.2 that after 62 days very low concentration increased height of the plants as compared to the high concentration. Number of leaves was same as compared to high concentration but as compared to control chitosan treatment increased number of leaves. Low concentration of chitosans increased number of flowers as compare to control. Table 21.3 showed that at the harvest time after 80 days in low concentration of chitosan height of the plant increased as compared to high concentration similarly leaves was found in low and in high concentration but chitosan treatment increased as compare to control. It is cleared from Table 21.4, 1 ml increase chitosans increased the leaf biomass as compared to high concentration and control while low concentration increased leaf area, and slightly increased number of fruit as well as the length of the fruit. It can be concluded that the chitosan treatment increased

Table 21.1: Effect of Chitosan on the Vegetative Growth of Okra (Bhindi) After 39 Days

Concentration (ml)	Height of the Plant (cm)	Number of Leaves
0	13	5
1	19	7
20	15.6	7
S.E	1.74	0.85
C.D	4.46	2.20

Table 21.2. Effect of Chitosans on the Vegetative Growth of Okra (Bhindi) at Flowering Stage After 62 Days

Concentration (ml)	Height of the Plant (cm)	Number of Leaves	Number of Flowers
0	21	5	1
1	47.3	7	3
20	41	7	2
S.E	7.93	0.53	0.39
C.D	20.38	1.37	1.01

Figure 21.1: Growth of the Chitosan Treated *Abelmoschus esculentus* Plants (Left control)

height of the plant, number of leaves, leaf area, number of fruit, length of the fruit and biomass as compare to control (Figure 21.1). These results are in confirmity with Ya Jing Guan *et al.* (2009) as well as Kananont *et al.* (2010). The similar effect was found in low and high concentration. Low concentration slightly increased the morphological parameters as compared to high concentration. High concentration did not produce the side effects on plant.

Table 21.3: Effect of Chitosans on the Vegetative Growth at the Time of Harvesting After 80 Days

Concentration (ml)	*Height of the Plant (cm)*	*Number of Leaves*
0	29	5
1	62	7
20	60	7
S.E	11.00	0.20
C.D	28.27	0.52

Table 21.4: Effect of Chitosans on the Component of Yield at the Time of Harvesting

Concentration (ml)	*Leaves Biomass*		*Leaf Area*	*Number of Fruits*	*Length of the Fruits (cm)*	*Weight of the Fruit (g)*
	Fresh	*Dry*				
0	0.910	0.130	0.35	1	12	10
1	2.590	0.630	0.94	3	16	19.6
20	1.680	0.390	0.72	2	15	19.6
S.E	0.49	0.14	0.17	0.49	1.20	1.20
C.D	1.25	0.37	0.44	1.26	3.09	3.09

References

Felter, Harvey Wickes and Lloyd, John Uri. (2007). *Hibiscus esculentus*–Okra *King's American Dispensatory*, 1898.

Kananont Nungruthai, Rath Pichyangkura, Sermsiri Chanprame, Supachitra Chadchawan and Patchra Limpanavech (2010). Chitosan specificity for the *in vitro* seed germination of two *Dendrobium* orchids (Asparagales: Orchidaceae). *Scientia Horticulturae*, **124(2)**: 239-247.

Khan, W.M., Prithiviraj, B. and Singh, D.L. (2002). Effect of foliar application of chitin oligosaccharides on photosynthesis of maize and soyabean. *Photosynthetica* **40**: 621-624.

Shao, C.X., Hu, J., Song, W.J. and Hu, W.M. (2005). Effect of seed priming with chitosan solution of different acidity on seed germination and physiological character in seediing. *Journal of Zhejiang University Agriculture and Life Science*, **31(6)**: 7.

Srinivasan, Rajani and Mishra, Anuradha (2008). Okra (*Hibiscus esculentus*) and fenugreek (*Trigonella foenum graecum*) mucilage characterization and application as flocllants for textile effluent treatments. *Chinese Journal of Polymer Science*, **26(6)**: 679-687.

Tantawy, E.M.E. (2009). Behaviour of tomato plants as affected by spraying with chitosan and aminofort as natural stimulator substances under application of soil organic amendments. *Pakistan Journal of Biological Science*, **12(11)**: 1164-1173.

Wojdyla, A.T. (2001). Chitosan in the control of rose disease-6 year trials. *Bull Polish Add Sci. Biol. Sci.*, **49**: 233-252.

Ya Jing Guan, Jin Hu, Xian Ju-Wang and Chen Xia Shao (2009). Seed priming with chitosan improves maize germination and seedling growth in relation to physiological changes under low temperature stress. *Journal of Zhejiang Univ. Sci. B*, **10(6)**: 427–433.

Chapter 22

Wetlands: Elixir of Life

S.N. Inamdar[1]

ABSTRACT

Wetlands are one of the most productive ecosystems, comparable to tropical evergreen forests in the biosphere and play a significant role in the ecological sustainability of a region. They are an essential part of human civilization meeting many crucial needs for life on earth such as drinking water, water purification, energy, fodder, biodiversity, flood storage, transport, recreation, research-education, sinks and climate stabilizers. The values of wetlands though overlapping, like the cultural, economic and ecological factors, are inseparable. Across the globe, they are getting extinct due to manifold reasons, including anthropogenic and natural processes. Burgeoning population, intensified human activity, unplanned development, absence of management structure, lack of proper legislation, and lack of awareness about the vital role played by these ecosystems (functions, values etc.) are the important causes that have contributed to their decline and extinction. With these, wetlands are permanently destroyed and loose any potential for rehabilitation. This has led to ecological disasters in some areas, in the form of large-scale devastations due to floods etc. So, there is a necessity to conserve these unique ecosystems for the future generation.

Keywords: Anthropogenic process, Productive ecosystems, Wetlands.

Introduction

Water is one of the major natural resources on earth, of which 97.5 per cent is salt water leaving only 2.5 per cent as fresh water of which, over 2/3rd is frozen in

---

1 Department of Botany, Miraj Mahavidyalaya, Miraj – 416 410, Dist. Sangli, Maharashtra, India

E-mail: snioasis@yahoo.com

glaciers and polar ice caps. Hence, there is an increasing demand for fresh water resources. Increase in population and its subsequent impacts further exert pressure on the fresh water demand, which needs the assessment of quality and quantity for its optimal utilization. Man still depends heavily on rainfall and running water. Industrial use of water leads to pollution. Insufficient planning and maintenance of water resources are also responsible for the wastage of the resources.

Since time immemorial wetlands have served as cradles of civilizations, the focal points around which great cultures took roots and flourished (Abbasi, 1997). Wetlands are estimated to occupy nearly 6.4 per cent of the earth's surface, 30 per cent of which is made up of bogs, 26 per cent fens, 20 per cent swamps, about 15 per cent flood plains, etc. (IUCN, 1989). The amount of fresh water on earth is very small compared to seawater, of which 69.6 per cent is locked away in the continental ice, 30.1 per cent in underground aquifers and 0.26 per cent in rivers and lakes. Lakes in particular occupy less than 0.007 per cent of the world's fresh water (Clarke, 1994).

Wetlands are among the most productive and biologically rich ecosystems on the earth (Richardson, 1995). They play a significant role in the ecological sustainability of a region. They are an essential part of human civilization meeting many crucial needs for life on earth such as drinking water, protein production, water purification, energy, fodder, biodiversity, flood storage, transport, recreation, research-education, sinks and climate stabilizers. The values of wetlands though overlapping, like the cultural, economic and ecological factors, are inseparable. The geomorphologic, climatic, hydrological and biotic diversity across continents has contributed to wetland diversity. Across the globe, they are getting extinct due to manifold reasons, including anthropogenic and natural processes. Burgeoning population, intensified human activity, unplanned development, absence of management structure, lack of proper legislation, and lack of awareness about the vital role played by these ecosystems (functions, values, etc.) are the important causes that have contributed to their decline and extinction. With these, wetlands are permanently destroyed and lose any potential for rehabilitation. This has led to ecological disasters in some areas, in the form of large-scale devastations due to floods etc.

Wetlands form the transitional zone between land and water, where saturation with water is the dominant factor determining the nature of soil development and the types of plant and animal communities living in and on it (Cowardin *et al.*, 1979). Wetland type is determined primarily by local hydrology and the unique pattern of water flow through an area. In general, there are two broad categories of wetlands: Coastal and Inland (Table 22.1).

Wetlands perform some useful functions in the maintenance of overall balance of nature. Some of the important functions are flood control, water storage and purification, protection of shorelines, floral and faunal habitats, gene pools, recreational besides providing outputs of commercial value and economic sustenance to the people.

Fortunately, during recent years, they have received a global attention, which started with the conference held in Ramsar (Iran) in 1971, where the first listing of wetlands of international importance was made. India as one of the original signatories has made impressive efforts in initiating work for conservation and management of wetlands.

Table 22.1: Broad Categories of Wetland

	Inland Wetlands	
1.	Natural	**Lakes/Ponds**
		Ox-bow lakes/Cut-off meanders
		Waterlogged (Seasonal) Playas
		Swamp/marsh
2.	Man-made	**Reservoirs**
		Tanks
		Waterlogged
		Abandoned quarries
		Ash pond/cooling pond
	Coastal Wetlands	
1.	Natural	**Estuary**
		Lagoon
		Creek
		Backwater (Kayal)
		Bay
		Tidal flat/Split/Bar
		Coral reef
		Rocky coast
		Mangrove forest
		Salt marsh/marsh
		Vegetation
		Other vegetation
2.	Man-made	Salt pan
		Aquaculture

India being developing country, wetlands disappear due to various anthropogenic developments. The major threats to wetlands include siltation, eutrophication, shrinkage of area, reclamation, encroachment, pollution, change in water quality, excessive tourism load, reduced arrival of migratory birds etc.

Definitions of Wetlands

Wetlands are ecotones or transitional zones that occupy an intermediate position between dry land and open water. Abundance of water at least for a part of the year is the single dominant factor. They possess characteristics of both terrestrial and aquatic ecosystems and properties that are uniquely of their own. Wetlands support a wide array of flora and fauna and deliver many ecological, climatic and societal functions. Scientists often refer to wetlands as the "kidneys" of the earth and forests as the "lungs" of the earth.

According to the Convention on Wetlands of International Importance held at Ramsar in Iran in 1971 and to which India is a signatory, Wetlands are defined as *"the areas of marsh, fen, peat land or water, whether natural or artificial, permanent or temporary, with water that is static or flowing, fresh, brackish or salty including areas of marine water, the depth of which at low tide does not exceed 6 meters. It may also incorporate riparian and coastal zones adjacent to the wetlands and islands or bodies of marine water deeper than 6 meters at low tide lying within the wetlands"*.

This definition emphasizes three key attributes of wetlands–(1) *hydrology* which is a degree of flooding or soil saturation, (2) *wetland vegetation* (hydrophytes) and (3) *hydric soils.*

According to Mitsch and Gosselink (1986), Wetlands are defined as *"lands transitional between terrestrial and aquatic ecosystems where the water table is usually at or near the surface or the land is covered by shallow water"*.

Thus, wetlands are areas of land where the water level remains near or above the surface of the ground for most of the year. As they support a variety of plant and animal life, biologically they are one of the most productive systems in the world.

Significance of Wetlands

Wetlands are one of the most productive ecosystems of the world; their primary productivity is known to exceed that of grasslands, cultivated lands and even tropical rain forests (Richardson, 1995).

Wetlands are often described as "kidneys of the landscape" (Mitsch and Gosselink, 1986). Hydrologic conditions can directly modify or change chemical and physical properties such as nutrient availability, degree of substrate assimilation, soil salinity, sediment properties and pH. These modifications of the physico-chemical environment, in turn, have a direct impact on the biotic response in the wetland (Gosselink and Turner, 1978).

Benefits of Wetlands

1. Nutrient cycling
2. Maintenance of biodiversity
3. Maintenance of water quality
4. Support commercial fishing, forestry
5. Water supply–Groundwater recharge
6. Habitat–Life support system for flora and fauna
7. Shoreline stabilization and erosion control
8. Winter resorts for a variety of birds for shelter and feed
9. Effective in flood control, wastewater treatment, reducing sediment load
10. Valuable for their educational and scientific interest (Diversity and species richness)

11. Recreational benefits (swimming, diving, hiking, fishing, hunting, bird watching, boating, tourism)
12. Aesthetic value

Threats Facing by Wetlands

Wetlands represent dynamic natural environments that are subjected to both human and natural forces. Natural events influencing wetlands include rising sea level, natural succession, hydrologic cycle, sedimentation and erosion. The rise in sea level, for example, both increases and decreases the wetland's spatial extent depending on local factors.

Wetlands are under increasing stress due to the rapidly growing population, technological development, urbanization and economic growth. Additional pressures on wetlands from natural causes like subsidence, drought, hurricanes, erosion etc., and human threats coming from over exploitation, encroachment, reclamation of vast wetland areas for agriculture, commercial and residential development, and silviculture have altered the rate and nature of wetland functions particularly in the last few decades. The primary pollutants causing degradation are Sediments, Nutrients, Pesticides, Salinity, Heavy metals, Weeds; Low dissolved oxygen, pH (Table 22.2).

Table 22.2: Contribution of Various Sectors to Pollution of Wetlands

Anthropogenic Activities	*Industrial*	*Domestic*	*Agriculture*	*Urbanization*
Discharges to Wetlands	✓	✓	✓	✓
Non-point Source Pollution	✓	✓	✓	✓
Air pollutants	✓	×	×	✓
Toxic chemicals	✓	✓	✓	×
Deposition of fills material	✓	✓	✓	✓
Construction	×	✓	×	✓
Tilling for crop production	×	×	✓	×
Pest species of plants and animals	×	✓	✓	×
Siltation	✓	✓	✓	✓
Changing nutrient levels	✓	✓	✓	×
Tourism and recreational activities	×	×	×	✓
Water regime and physical modification	✓	✓	✓	✓

About 50 per cent of the world's wetlands have been lost in the last century, primarily through drainage for agriculture, urban development and water system regulations.

On a global scale, climate change could also affect wetlands through increased air temperature; shifts in precipitation; increased frequency of storms, droughts, and floods; increased atmospheric carbon dioxide concentration; and sea level rise. Apart from pollution, the other major problems include hydrologic manipulations of wetlands in the form of flow alterations and diversions, disposal of dredged or fill material, sewage inflows leading to alterations in:

1. Water currents, erosion or sedimentation patterns
2. Natural water temperature variations
3. Chemical, nutrient and dissolved oxygen regime of the wetland
4. Normal movement of aquatic fauna
5. pH of the wetland and
6. Normal water levels or elevations.

All of these impacts affect the wetland quality, species composition and functions.

Threats to Wetland Ecosystems–Biotic and Abiotic

Biotic

1. Uncontrolled siltation and weed infestation.
2. Uncontrolled discharge of wastewater, industrial effluents, surface run-off, etc. resulting in proliferation of aquatic weeds, which adversely affect the flora and fauna.
3. Tree felling for fuel wood and wood products causes soil loss affecting rainfall pattern, loss of various aquatic species due to water-level fluctuation.
4. Habitat destruction leading to loss of fish and decrease in number of migratory birds.

Abiotic

1. Encroachment resulting in shrinkage of area.
2. Anthropogenic pressures resulting in habitat destruction and loss of biodiversity.
3. Uncontrolled dredging resulting in successional changes.
4. Hydrological intervention resulting in loss of aquifers.
5. Pollution from point and non-point sources resulting in deterioration of water quality.
6. Ill-effects of fertilizers and insecticides used in adjoining agricultural fields.

Major Problems for Wetland Conservation

1. Lack of awareness of wetland values and functions
2. Lack of information and data on wetlands
3. Lack of integrated land use plan
4. Inappropriate water use planning and implementation
5. Unsustainable use of wetlands

Desirable Actions to Manage the Wetlands

1. Mapping
2. Baseline data
3. System approach
4. Control of construction activities
5. Constant water quality monitoring
6. Afforestation in catchment area
7. Awareness building
8. Setting up of a separate authority
9. NGO's involvement
10. Education

Summary and Conclusion

Across the globe, the wetlands are getting extinct due to manifold reasons, including anthropogenic and natural processes. Burgeoning population, intensified human activity, unplanned development, absence of management structure, lack of proper legislation, and lack of awareness about the vital role played by these ecosystems (functions, values etc.) are the important causes that have contributed to their decline and extinction. With these, wetlands are permanently destroyed and loose any potential for rehabilitation. There is obviously much ground to be covered in our conservation efforts of wetlands and it is the strong need of the hour. Since wetlands are a common property resource, it is an uphill task to protect or conserve these ecosystems for the future generation.

References

Abbasi, S.A. (1997). *Wetlands of India–Ecology and Treats: The Ecology and the Exploitation of Typical South Indian Wetlands*, Vol. I. Discovery Publ. House, New Delhi.

Clarke, R. (1994). *The Pollution of Lakes and Reservoirs*. UNEP Environment Library no. 12, Nairobi, Kenya.

Cowardin, L.M., Cartor, V., Golet, F.C. and Laroa, E.T. (1979). Classification of Wetlands and Deep Habitats of the United States. U.S. Fish and Wildlife Service FWS/OBS-79/31. Washington D.C. Prescott, G. W. 1982. Algae of the Western Great Lakes Areas, Otto. Koeltz Science Publ., Germany, pp. 662-962.

Gosselink, J.G. and Turner, R.E. (1978). The role of hydrology in fresh water wetland ecosystems. In: *Freshwater Wetlands: Ecological Processes and Management Potential*, (Eds.) R.E. Good, D.F. Whigham and R.L. Simpson. Academic Press, New York, pp. 63-78.

Mitsch, W.J. and Gosselink, J.G. (1986). *Wetlands*. Van Nostrand Reinhold, New York.

Richardson, C.J. (1995). Wetlands ecology. In: *Encyclopedia of Environmental Biology*, (Ed.) W.A. Niereburg. Academic Press, New York.

IUCN (1989). *A Directory of Asian Wetlands*.

Chapter 23

Chemical Control of Fruit Rot of *Coccinia indica* Caused by *Bipolaris tetramera* and *Macrophomina phaseolina*

*V.S. Chatage[1], A.S. Sonwane[2] and U.N. Bhale[1]**

ABSTRACT

Ivy gourd (*Coccinia indicia* Wight and Arn.) suffers from pre and postharvest diseases. This paper describes the sensitivity (MIC) of *Bipolaris tetramera* and *Macrohomina phaseolina* against mancozeb and carbendazim fungicides. Sensitivity of 10 isolates of *B. tetramera* was tested against mancozeb, there was large variation in the sensitivity of these isolates. Some isolates were sensitive (20µg/ml) while others were resistant (120µg/ml) *i.e.* ranged from 20 to 120µg/ml while in case of carbendazim sensitivity ranged from 1800 to 4000 µg/ml. In *M.phaseolina* the sensitivity of 10 isolates were tested against mancozeb and carbendazim. In mancozeb some isolates were sensitive and resistant *i.e.*ranged from 100 to 800µg/ml while in case of carbendazim sensitivity ranged from 2500 to 4900µg/ml

Keywords: *Coccinia indica, Bipolaris tetramera, Fungicides Macrohomina phaseolina etc.*

---

1 Research Laboratory, Department of Botany, Arts, Science and Commerce College, Naldurg Tq. Tuljapur, Dist. Osmanabad – 413 602, M.S., India

2 Department of Botany, Shadabai Pawar Mahavidyalaya, Baramati Dist Pune, M.S., India

* Corresponding Author E-mail: unbhale2007@rediffmail.com

Introduction

IVY gourd (*Coccinia indicia* Wight and Arn.) of family Cucurbitaceae is distributed in tropical Asia, Africa, Pakistan, and India and Sri Lanka (Cook 1903, Sastri 1950). It is a climber and trailer (Nasir and Ali, 1973). The fruit is used as vegetable when green and eaten fresh when ripened. Every part of this plant is valuable in medicine. The IVY gourd, however suffers from pre and post harvest disease caused by *B.tetramera* and *M. phaseolina*.The cultures were deposited at ASC College Naldurg. Investigation was undertaken to evaluate the sensitivity of *B.tetramera, M.phaseolina* against mancozeb and carbendazim fungcides.

Material and Methods

Sensitivity of isolates were tested against mancozeb and carbendazim determined by food poisoning test (Dekker and Gielink, 1979). Czapek Dox agar plates were prepared containing different concentration of fungicides mancozeb and carbendazim.Seven days fresh culture disc (8mm) of the isolates were inoculated at the centre of plates in triplicate. The plates were incubated at 28±°C in the dark and radial growth was measured at different intervals. Plates without fungicides treated as control. *In vitro* studies, *B. tetramera* for mancozeb (10 to120 µg/ml) and for carbendazim (1100 to 4000 µg/ml) concentrations were prepared. While in *M.phaseolina* for mancozeb (100 to 800 µg/ml) and carbendazim (1000 to 4900µg/ml) were prepared.

In *vivo* experiment was conducted on ivy gourd fruits by using the different concentrations ranged from 10 to 220µg/ml and carbendazim ranged 2000 to 4500 µg/ml for *B. tetramera, M. phaseolina* for mancozeb for ranged 100 to 900 µg/ml. from carbendazim 2000 to 5000 µg/ml ranged were used.

Results and Discussion

This paper describes the sensitivity (MIC) of *B. tetramera* and *M. phaseolina* against mancozeb and carbendazim fungicides Sensitivity of 10 isolates of *B. tetramera* was tested against mancozeb. There was large variation in the sensitivity of these isolates. Some isolates were sensitive (20µg/ml) *i.e.* Bt_2 while others were resistant (120µg/ml). *i.e.* Bt_5, Bt_6, Bt_8, Bt_9, Bt_{10}. Its rainged from (10 to 120 µg/ml). In case of carbendazim, sensitivity ranged from (2000 to 4000 µg/ml) and sensitive isolates are Bt_1, Bt_2, Bt_3, Bt_4, Bt_7, Bt_8 and Bt_9 and tolerant isolates were Bt_5, and Bt_{10} (Table 23.1).

Table 23.2 showed that the sensitivity of *M.phaseolina* 10 isolates in mancozeb, some isolates were sensitive isolates Mp_1 (100 µg/ml) and highly tolerant isolates Mp_9(800 µg/ml) while in case of carbendazim, sensitivity ranged from 3000 to 4900 µg/ml. In *vivo* experiments, showed positive results. Sensitivity of *B. tetramera* against mancozeb ranged from 10 to 220 µg/ml while in carbendazim its ranged from 2000 to 4500 µg/ml. In case of *M. phaseolina* against mancozeb ranged from 100 to 900 µg/ml while in carbendazim it was 2000 to 5000 µg/ml.

Variation in the sensitivity of different pathogens in relation to many fungicides have been reported (Jones and Ehret, 1976, Dekker and Gielink, 1979, Gangawane and Saler, 1981; Gangawane and Shaikh, 1988; Hollomon, 1981, Kamble, 1991, Bhale,

Table 23.1: Sensitivity of *Bipolaris tetramera* Isolates from *Coccinia indica* Against Fungicide

Sl.No.	*Isolate*	*Place*	*Fungicides*			
			Mancozeb (MIC) μg/ml (9mm disc)		*Carbendazim (MIC) μg/ml (9mm disc)*	
			In vitro	*In vivo*	*In vitro*	*In vivo*
1.	Bt_1	Murum	30	40	1900	2000
2.	Bt_2	Solapur	20*	22*	2800	2900
3.	Bt_3	Pune	45	46	1900	2000
4.	Bt_4	Mumbai	35	37	2700	2800
5.	Bt_5	Jewali	70	75	3500	3600
6.	Bt_6	Latur	80	90	4000+	4500+
7.	Bt_7	Naldurg	35	40	2900	3000
8.	Bt_8	Omerga	60	70	1800*	2000*
9.	Bt_9	Aurangabad	120+	220+	2500	2700
10.	Bt_{10}	Haralee	100	200	3000	3100

* Sensitive + Resistant; Bt: *Bipolaris tetramera*; MIC: Minimum Inhibitory Concentration.

Table 23.2: Sensitivity of *Macrohomina phaseolina* Isolates from *Coccinia indica* Against Fungicide

Sl.No.	*Isolate*	*Place*	*Fungicides*			
			Mancozeb (MIC) μg/ml (9mm disc)		*Carbendazim (MIC) μg/ml (9mm disc)*	
			In vitro	*In vivo*	*In vitro*	*In vivo*
1.	Mp_1	Aurangabad	100*	120*	4400	4500
2.	Mp_2	Beed	200	300	2800	3000
3.	Mp3	Jalna	230	350	4800	5000
4.	Mp_4	Pune	150	200	4200	4400
5.	Mp_5	Nilanga	250	350	4800	4900
6.	Mp_6	Thane	120	200	3000	3100
7.	Mp_7	Latur	500	600	3500	3600
8.	Mp_8	Solapur	600	700	4900+	5000+
9.	Mp_9	Osmanbad	800+	900+	4000	4100
10.	Mp_{10}	Nasnik	700	800	2500*	2700 *

* Sensitive + Resistant; Mp: *Macrohomina phaseolina*; MIC: Minimum Inhibitory Concentration.

2002). Annamalai and Lalithakumari (1996) suggested that it is essential to establish the base line sensitivity for the fungicide against sensitive strain, Brain (1980) considers that heterogeneous population of nuclei consisting of resistant and sensitive nuclei in the isolates might be responsible for variation in the MIC of fungicides. Recently Bhale and Gogle reported the development of carbendazim resistance in *Alternaria spinaciae* incitant of spinach (*spinacia oleracea L.*). There was variation in MIC of Ridomil Gold among the five isolates of *Phytophthora palmivora* var. *piperina* on the agar plates (Patil and Kamble, 2011).

References

Annamalai, P. and Lalithakumari, D. (1996). Decreased Sensitivity of *Dreechslera oryzae.* Field isolates to edifenphos, *Ind. Phytopath.*, **43(4)**: 553-558.

Bhale, U.N. (2002). Studies on management of some important diseases of spinach in Maharashtra. *Ph.D. Thesis,* Dr. Babasaheb Ambedkar Marathwada University, Aurangabad.

Bhale, U.N. and Gogle, D.P. (2008). Effect of passage on the Development of cabendazim Resistance in *Alteraria spinaciae* incitant of leaf spot of spinach (*Spinacea oleracea* L.). *Geobios,* **35**: 37-40.

Brain, G.C.A. (1980). Resistance in perenospo rale to acylalanine type fungicides. *Ph.D. Thesis,* Univ. Guelph, Ontario, Canada, pp. 10.

Cooke, C.I.E.T. (1903). *Flora of Presidency of Bombay,* Vol 1. Published under the Authority of Secretary of state for Council.

Dekker, J. and Gielink, A.J. (1979). Acquired resistance to pimaricin in *Cladosporium cucumerinum* and *Fusarium oxysporumf.* sp. *narcissi* associated with decreased virulence. *Neith. J. Plant Pathol.* **85**: 67-73.

Gangawane, L.V. and Saler, R.S. (1981). Resistance to fungicides in *Aspergillus flavus. Neth. J. Pl.Path.* **87**-254.

Gangawane, L.V. and Shaikh, S.A. (1988). Development of resistance to aluminium ethyl phosphate in *pythium aphani dermatum. Indian Phytopathology,* **41(41)**: 638-641.

Hollomon, D.W. (1981). Mode of action of hadroxy pyrimidine fungicides. In: *Fungicides Resistance in Protection,* (Eds.) J. Dekker and S.G. Geogopoulous. CAPD Wegeningen, Netherlands.

Jones, A.L. and Ehret, G.R. (1976). Tolerance to fungicides in *Venturia* and *Monilinia* of tree fruits. *Proc. Ann. Phytopath. Soc.,* **3**: 84-90.

Nasir, E. and Ail, S.I. (1973). *Flora of West Pakistan, Cucurbitaceae,* No. 154, Botany Department, University of Karachi.

Patil, V.B. and Kamble, S.S. (2011). Efficacy of ridomil gold against *Phytophthora palmivora* var. *piperina* causing quick wilt of black pepper. *Bioinfolet,* **8 (1)**: 105-106.

Sastri, B.N. (1950). *The Wealth of India: A Dictionary of Raw Materials and Industrial Products.* Publication and information of Directorate CSIR, New Delhi, **2&8**: 257, 285-293.

Chapter 24

Antibiotic Effect of Culture Filtrates of *Fusarium oxysporum* f.sp. *udum* Butler Against *Rhizobium* Bacteria Isolated from Different Varieties of Pigeonpea (*Cajanus cajan* L. Millsp.)

V. Jalander[1] and B.D. Gachande[1]*

ABSTRACT

Rhizobium bacteria were isolated from ten different varieties of pigeonpea (*Cajanus cajan* L. Millsp.) root nodules on yeast extract mannitol agar (YEMA) medium and they were tested for their sensitivity to metabolites (culture filtrates) of *Fusarium oxysporum* f.sp. *udum* Butler isolated from seven different pigeon pea varieties on YEMA medium. All the ten isolates of *Rhizobium* bacteria were sensitive

1 Botany Research Laboratory and Plant Disease Clinic, N.E.S. Science College, Nanded, M.S. India

* Corresponding Author E-mail: jalandervaghmare@gmail.com

to all the pathogenic *Fusrium* isolates. Bacterium isolated from var. PUSA-992 was highly sensitive to metabolites of *F. oxysporum* f.sp. *udum* isolated from ICPL-87119. The bacterium isolated from var. BSMR-853 was highly resistant to culture filtrate of the pathogen isolated from the var. ICP-2376 when compared with other isolates.

Keywords: Antibiosis, Culture filtrates, F. oxysporum udum, Pigeonpea, Rhizobium.

Introduction

Red Gram also known as Pigeon pea (Arhar or tur in local language) is an important pulse crop of India, and is being cultivated on 35.6 lakh ha area. Among total pulses, the red gram accounts for 14.5 per cent in area and 15.5 per cent in productivity. Maharashtra is the largest producer with approximately 10.51 lakh ha area with average productivity of 6.03 Q/ha. Being important nitrogen fixing crop, it is widely grown for enriching the soil. Its deep penetrating roots helps in bringing nutrients from deeper layers of soil. Falling of leaves before maturity ensures sufficient incorporation of nitrogen and other nutrients in the soil, which are beneficial to the subsequent crop. Nitrogen fixing potential of red gram makes it an ideal intercrop/ rotation crop in organic management. Subsequent crops taken after red gram are reported to benefit by up to 30 kg N/ha. Red gram is grown mainly as intercrop between sorghum (Jowar), pearl millet (Bajra), Maize and cotton. Under irrigated conditions it can also be cultivated as monocrop.

The roots of pigeonpea plants growing in soil provide a unique habitat, the rhizosphere which is particularly favorable for the development of soil-microorganisms. A large number of microorganisms are known to produce toxic metabolites when cultivated on synthetic media. Fungal metabolites are substances discharged by fungi in their metabolic processes. The metabolites are products of some aminoacids, cyclic peptides, aromatic phenols, terpinoids and plant growth regulators (Graffin, 1981; Madhosingh, 1995; Nema, 1992).

Materials and Methods

Isolation of *Rhizobium* Bacteria

Ten different varieties of pigeon pea (*Cajanus cajan* L. Millsp.) *viz.* PUSA-992, BDN-2, BDN-708, BSMR-736, BSMR-853, BSMR-175, ICPL-87119, ICP-8863, ICP-2376 and AKT-9913 were selected for the study. Root nodules of the above varieties were collected in sterile polythene bags from Pulses research center, Badnapur, Dist.Jalna (M.S.) and transferred to laboratory in a ice cooled container having 4 to 6°C temperature. Isolation of *Rhizobium* bacteria was done on yeast extract mannitol agar (YEMA) medium (Aneja, 2007). Pure cultures were maintained on same medium slants and stored in refrigerator (4 ± 2°C) for further use.

Collection of Diseased Root Samples and Isolation of *Fusarium oxysporum* f.sp. *udum*

The infected root samples of host plant varieties were collected in polythene bags from the same center (above mentioned) in the presence of expert person of the center and transferred to laboratory. Isolation of *F. oxysporum* f.sp. *udum* was done on czapek's dox agar (CDA) medium by tissue segment method and then pure cultures of the pathogen was maintained.

Collection of Culture Filtrates

The seven different isolates of *Fusarium oxysporum udum* were grown in 250 ml conical flask containing 100 ml Czapek's dextrose liquid medium for ten days at 25±2°C in BOD incubator. After incubation culture filtrates were filtered in pre sterilized flasks by using Whatman no.50 filter paper and stored at 4±1°C in refrigerator.

Antibiotic Sensitivity Test

Antibiotic sensitivity test was carried out by agar well diffusion method (Perez *et al.*, 1990). YEM agar medium plates were prepared and shredded with the inoculums of different bacterial isolates then 5 mm diameter well was made with the help of sterile cork borer. The well was filled with 0.5ml of culture filtrate of the pathogen and then plates were incubated for 48 hours at room temperature. Three replicates were maintained for each isolate. After incubation plates were observed and the zone of inhibition was measured (mean of three replicates).

Table 24.1: Antibiotic Effect of Culture Filtrates of *Fusarium oxysporum* f.sp. *udum* Butler Against *Rhizobium* Bacteria Isolated from Different Varieties of Pigeonpea (*Cajanus cajan* L. Millsp.)

Pigeon Pea Varieties	*Zone of Inhibition in cm*							
	PUSA-992	*BDN-2*	*BDN-708*	*ICPL-87119*	*ICP-8863*	*ICP-2376*	*AKT-9913*	*Control*
PUSA-992	2.2	2.7	2.5	**4.0**	3.0	2.5	3.2	2.6
BDN-2	2.3	2.5	2.3	2.5	2.4	2.8	2.4	3.0
BDN-708	1.5	2.0	2.2	2.4	2.3	2.0	2.6	3.0
BSMR-853	1.6	2.6	2.0	1.8	1.5	**1.2**	1.4	2.2
BSMR-736	2.2	2.5	2.3	2.5	2.8	1.8	2.2	2.5
BSMR-175	1.5	2.3	1.7	2.0	1.7	1.5	1.8	2.5
ICPL-87119	2.0	1.5	1.5	2.2	1.5	1.5	1.5	3.5
ICP-8863	2.0	1.5	1.7	2.4	1.8	1.6	2.0	2.8
ICP-2376	1.7	1.5	2.0	2.4	1.7	1.5	1.5	2.6
AKT-9913	2.0	2.3	1.5	2.5	2.0	1.5	1.8	2.8

Results and Discussion

Results were presented in Table 24.1 showed that the isolates of *F. oxysporum* f.sp. *udum* isolated from different varieties of pigeon pea were inhibited the growth of *Rhizobium* isolates of different varieties of pigeon pea. The bacteria isolated from vars. PUSA-992, ICPL-87119, ICP-8863, ICP-2376 and AKT-9913 were highly sensitive to *F. oxysporum* f.sp. *udum* isolated from variety ICPL-87119 and the zone of inhibition 4.2, 2.55, 2.44, 2.44 and 2.55 cm respectively. The bacterium isolated from variety

BSMR-853 and 175 were sensitive to pathogen isolated from var. BDN-2. The remaining isolates of *Rhizobium* bacteria of var. BDN-2 were sensitive to pathogen of var. ICP-2376, bacteria of BDN-708 to var. AKT-9913 and bacteria of var. BSMR-736 highly sensitive to pathogen isolated from var. ICP-8863. The bacterium isolated from var. BSMR-853 was highly resistant to the pathogen isolated from ICP-2376. The above results conformed to the findings of Chaitanya *et al.* (2008), who were observed inhibitory effect of culture filtrates of rhizosphere fungi isolated from green gram. Similar results were also reported by Anusuya and Sullia (1984) in case of some common soil fungal metabolites against four *Rhizobium* strains isolated from groundnut root nodules. Nemec *et al.* (1964) were observed the antibiotic activity of soil fungal culture filtrates against some bacteria and yeasts.

References

Aneja, K.R. (2007). *Experiments in Microbiology, Plant Pathology and Biotechnology.* New Age International (P) Ltd. Publishers, New Delhi.

Anusuya, D. and Sullia, B. (1984). The antibiotic effect of culture filtrates of some fungi on rhizobial growth in cultures. *Plant and Soil*, **77(2-3)**: 387-390.

Chaitanya, C., Sridevi, M. and Mallaiah, K.V. (2008). Effect of rhizosphere fungi on *Rhizobium* and nodulation of green gram (*Vigna radiate* L. Wilezek). *The ICFAI Univ. J. of Life Sciences*, **2(3)**: 42-50.

Griffin, D.H. (1981). *Fungal Physiology*. John Wiley and Sons, New York, pp. 383.

Madhosing, C. (1995). Relative wilt-inducing capacity of the culture filtrates of isolates of *Fusarium oxysporum* f.sp. *radicis-lycopersici*, the tomato crown and root-rot pathogen. *J. Phytopathol.*, **4**: 193-198.

Nema, A.G. (1992). Studies on pectinolytic and cellulolytic enzymes produced by *Fusarium udum* causing wilt of Pigeonpea. *Indian J. Forest*, **15**: 353-355.

Nemec, P., Barath, Z., Betina, V. and Marta Kufkova (1964). Antibiotic activity of fungi isolated from soil samples from Indonasia. *Folia Microbiologia*, **9(6)**: 383-386.

Perez, C., Paul, M. and Bazerque, P. (1990). Antibiotic assay by well diffusion method. *Acta Biol. Med. Expt.*, **15**: 113-115.

Chapter 25

Aeschynomene villosa Poir (Fabaceae): An Addition to the Flora of Maharashtra State

S.P. Gaikwad[1], K.U. Garad[1] and R.D. Gore[1]*

ABSTRACT

Present paper reports *Aeschynomene villosa* Poir for the first time to the State of Maharashtra. It occurred on the wet margins of water canals and bunds in GIB wildlife sanctuary, Nannaj in Solapur district of Maharashtra. Brief description, habitat, ecology, phonological and GPS data, and photographs are given for easy identification of the species.

Keywords: *Aeschynomene villosa Poir, New record, Maharashtra state.*

Introduction

Genus *Aeschynomene* L. is represented by *c.* 150 species mainly distributed in the warmer region of the World (C.D.K. Cooke, 1996). In India, earlier the genus was known by three species *viz. A. americana* L., *A. aspera* L. and *A. indica* L. (Sanjappa, 1991), however, recently Dave (2004) has reported *A. villosa* Poir from Gujarat State as new record for India. So that presently total four species of *Aeschynomene* are found in

1 Life Science Research Laboratory, Walchand College of Arts and Science, Solapur – 6, Maharashtra, India

* Corresponding Author E-mail: sayajiraog@gmail.com

Plate 25.1: ***Aeschynomene villosa*** **poir**

India. In the State of Maharashtra, the genus is represented by three species. (Cooke, 1958, Almeida, 1998, Naik, 1998, Kothari, 2000).

During a floristic exploration in Great Indian Bustard (GIB) wildlife sanctuary, Nannaj, Maharashtra, we came across population of the *Aeschynomene*, which did not match with species found in Maharashtra. A perusal of relevant literature and critical examination revealed that the collected specimens belong to *Aeschynomene villosa* Poir. Identification was confirmed at Botanical survey of India, Calcutta. It is reported in this paper as new record for Maharashtra State. The voucher specimens are deposited in the Herbarium of department of Botany, Shivaji University, Kolhapur and BSI, Pune, Maharashtra.

Taxonomic Account

Aeschynomene villosa Poir. in Lam., Encyc. Suppl. 4: 76. 1816; Rudd, Contr. U.S. Natl. Herb. 32: 32.1955 and Reinwardtia 5: 27. 1959; Verdcourt, Man. New Guinea Leg. 368. 1979. Type: Puerto Rico, savannas, *Ledru s.n.* (P,*n.v.*). (Plate 25.1).

Erect herbs, up to 1.5 m tall; stems hispid; stipules peltate-appendiculate, hispid, especially at the point of attachment, striate, ciliate, 5-10 mm long, 1-1.5 mm wide, the upper portion attenuate, slightly longer than the lower attenuate or erose portion. Leaves about 2-7 cm long, 20-50 foliate, petiole and rachis glandular hispidulous; leaflets 4-15 mm long, 1-3 mm wide, apiculate, base asymmetrically rounded, glabrous, sub-falcate, 2 to several costate, ciliate; racemes axillary 3-7 flowered, rachis hispidulous; bracts cordate, 1.5-6 mm long, 1-2 mm wide, ciliate, acuminate; bracteoles ovate-lanceolate, 1-4 mm long, 0.5-1 mm wide, ciliate, acute to acuminate, flowers 3-6 mm long, calyx hispidulous, bilabiate, 3-5 mm long, petals pinkish, glabrous; stamens 10, diadalphous, 5+5; anthers uniform, versatile; style glabrous, with a minute terminal stigma. Fruits 1-3 cm long, 3-9 seeded, stipe 1.5-2 mm long, articulations weak, articles sub orbicular, 2-3 mm in diameter, glandular villous, venation inconspicuous, margins often breaking away from the body of articles; seeds 2-2.5 mm long, 1.5-2 mm wide, blackish.

Flowering and Fruiting: August–November.

Habitat and Ecology: *Aeschynomene villosa* Poir was occasionally found growing on the poorly drained wet margins of water canals and bunds of cultivated fields at 1739 ft. above mean sea level in association with *Achyranthes aspera* L., *Cassia obtusifolia* L., *Peristrophe paniculata* (Forssk.) Brummit etc.

GPS Data: N17 49.098 E75 58.068 at 1739 ft from above mean sea level.

Distribution: Tropical and subtropical America, introduced in India, Sri Lanka, Malaysia, Australia, New Guinea.

Acknowledgement

Authors are thankful to the Principal, Walchand College of Arts and Science, Solapur for providing available research facilities; Director, Botanical Survey of India, Calcutta for confirmation of identifications and to RGSTC, Govt. of Maharashtra for financial assistance.

References

Almeida, M.R. (1998). *Flora of Maharashtra.* Blatter herbarium, St. Xavier's College, Mumbai, **2**: 5-6.

Cooke, C.D.K. (1996). *Aquatic and Wetland Plants of India.* Oxford University Press, pp. 206-208.

Cooke, T. (1958). *Flora of Bombay Presidency* (Repr. ed.). Govt. of India, Bot. Surv. India 1: 362-363.

Dave, Mona (2004). New plant records for India. *Rheedea,* **14 (1&2)**: 61-62

Naik, V.N. (1998). *Flora of Marathwada.* Amrut Prakashan, Aurangabad, **1**: 241-242.

Sanjappa, M. (1991). *Legumes of India.* Bishen Singh and Mahendra Pal Singh, Dehradun, India, pp. 75.

Chapter 26

Diversity of Fungal Species Found in Different College Libraries of Wardha City During the Rainy Season

L.P. Dalal[1], M. Bhowal[2*] and S.P. Kalbende[2**]

ABSTRACT

The present work deals with the investigation, isolation and study of monthly variation of airborne fungal flora from different college libraries of Wardha city (M.S.). Aeromycoflora from the libraries was studied by using the exposed culture plate method by exposing petri plates with PDA and CZA medium to air for 10 minutes and then counting the number of growing colonies. This survey was carried out during the rainy season. Isolated fungi were identified and classified on the basis of cultural and morphological characters with the help of authentic literature. These includes fungi such as *Aspergillus flavus, Alternaria, Curvularia* etc. The cellulose degrading fungi such as *Rhizopus, Aspergillus niger, Penicillium* were also identified.

Keywords: *Aeromycoflora, Cellulose degrading fungi, CZA, PDA.*

---

1 Department of Botany, J.B. College of Science, Wardha, M.S., India

2 Department of Botany, Hislop College, Nagpur, M.S., India

E-mail: *mousumi_bhowal@rediffmail.com, **kalbendeswapna@gmail.com

Introduction

Fungal spores, pollen grains, and some bacteria are among the most prevalent airborne particles which form the major components of our environment but also act as biopollutants. The available literature (Sreeramulu, 1967) shows that airspora studies have been done almost exclusively in outdoor environments. However, human exposures also occur to a significant extent indoors. The present research considers few specific indoor environments namely, libraries since there has been little planned scientific study with respect to the air spora of libraries. On these issue extensive study have been carried out abroad. In India such work was carried out in past by many workers like Tilak and Vishwe (1975), Tilak and Chakre (1978), Babu (1983), Tilak and Pillai (1988), Sinha *et al.* (1998) etc. Recently only few records were seen highlighting indoor like Verma *et al.* (2004), Tiwari (2005), Pillai and Patil (2007), Patil *et al.* (2009).

Biodeterioration of library materials is a worldwide problem and it causes great damage especially to unique manuscripts and books stored in the libraries (Zyska, 1993). In this study we intend to discuss one of the biological factors as a main external group of factors that influence library materials. Fungal spores are an important component of the bioaerosol. There are about 80,000 species most of which are cosmopolitan in origin. The biological features of the fungi *i.e.* their ease of dispersion makes fungi one of the chief agents of contamination of any type of substrate including cellulose materials in the books of library. Many are pathogenic to human beings causing allergic problems including asthma due to differential deposition in the respiratory system. These fungi along with bacteria are responsible for the deterioration of the materials in the library. The activity of different environmental factors may cause some changes in physical and chemical properties of library collections and most of the time it has been seen these conditions are conducive for the growth of microbes hence, accelerating the deterioration process. In addition to internal causes of the deterioration of paper in books, due to its acidity, external agents are also a major threat to manuscripts (Wessel, 1970; Zyska, 1993).

Materials and Methods

Isolation of aeromycoflora from the different college libraries was carried out during rainy season by using the exposed culture plate method. The petridish culture plate method was used for determination of total number of colonies recorded in the entire sampling period. The sites was sampled every month with petriplates of 10cm diameter containing two different media viz; Czapek Dox Agar (CZA) and Potato Dextrose Agar (PDA) with dissolved streptomycin to exclude bacterial colonies. These petriplates were exposed at the height of one meter above the ground level for 10 min. and then incubated at 28 ± 1°C. for 4-5 days and were regularly examined. After few days the petriplates were observed for number and distribution of fungal colonies on agar plates and recorded in the notebook for total number of colonies, species present. These fungal colonies were isolated and cultured on appropriate media for identification and further studies. For species identification, specimen microscopic slides were prepared using glycerine jelly as mounting media and lactophenol cotton blue as the standard stain. Record of temperature, humidity etc. were also maintained. The fungal species were identified using the available literature.

Results and Discussions

The results of aeromycoflora during rainy season showed that the indoor library atmosphere was never free from fungal spores. A total of 11 types of fungal species were identified up to generic level and some of them are up to species level and the rest which are not identified were grouped under unidentified fungal types (Table 26.1). Among the isolated fungi some are as shown in Figure 26.2. Airborne fungal spores recorded were representatives from the three major groups *i.e.* Ascomycotina, Zygomycotina and Deuteromycotina. The monthly variation pattern revealed that *Aspergillus* species were dominant in all the libraries during rainy season followed by *Alternaria, Penicillium, Curvularia* etc. Out of all other species of *Aspergillus* frequency of *Aspergillus niger* dominated followed by *A. flavus, A. fumigatus* etc. (Figure 26.1) and the least dominant the *Trichoderma* followed by *Fusarium*. One of the reason might be because these spores have the capability to multiply faster than the rest. Along with the petriplates exposure method the fungi were also isolated from the books in libraries which shows similarity in both, the fungi isolated from the air and from the books. Among the isolated fungi from books, colonies of *Aspergillus, Penicillium* and *Rhizopus* had contaminated the books which are also isolated by the petriplate exposure method in rainy season.

Table 26.1: Fungal Species Identified Using Petriplates Exposure Method

Fungal Species	*College Libraries*				
	B.D. College of Engineering	*G.S. Commerce College*	*Yeshwant Arts College*	*J.B. Science College*	*R.G. Bhoyar Group of Institution*
Aspergillus niger	+	+	+	+	+
Aspergillus flavus	+	+	+	–	–
Aspergillus fumigatus	–	–	+	+	–
Alternaria alternata	+	+	–	+	+
Alternaria spp	+	–	+	+	+
Rhizopus spp	–	–	+	+	–
Curvularia spp.	+	+	–	+	+
Penicillium spp.	+	+	–	+	–
Helminthosporium spp.	+	+	–	+	–
Trichoderma spp.	–	–	–	–	+
Fusarium spp.	–	–	+	–	–
Unidentified spp.	+	+	+	+	+

+: Present; –: Absent.

Biodeteriogens like fungi and bacteria causes severe damage to library materials. Microbes get entry into indoor atmosphere through wind currents and settle on various

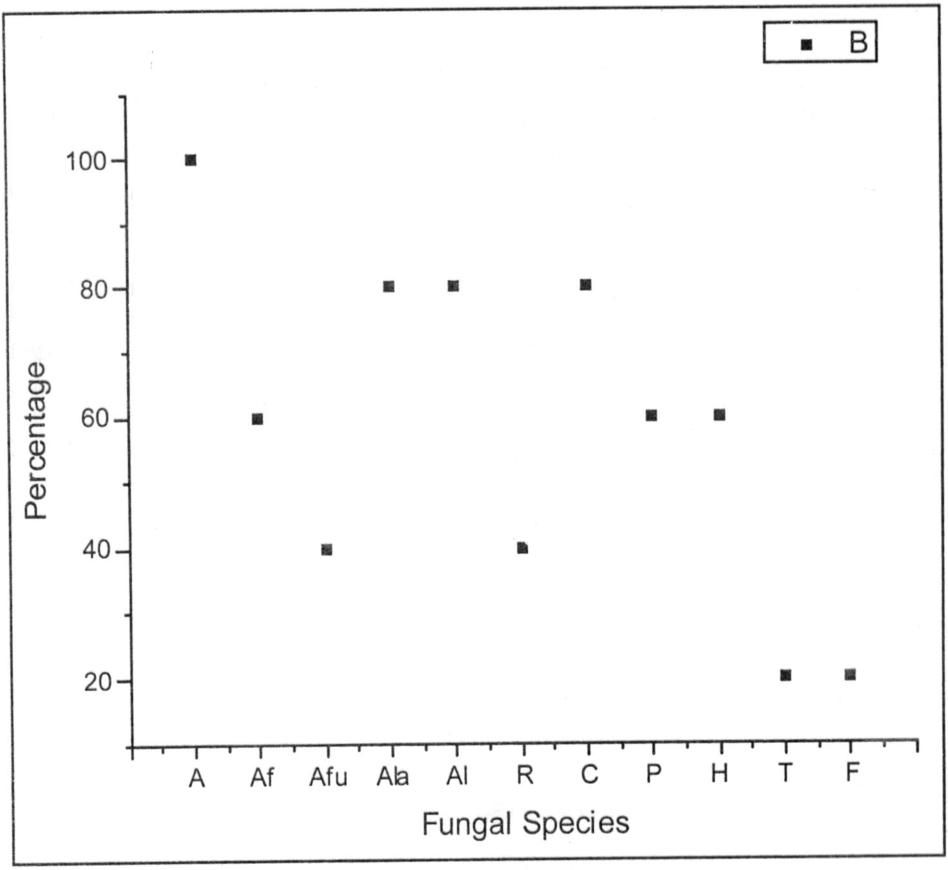

Figure 26.1: Concentration of Fungal Species during Rainy Season

A: *Aspergillus niger*; Af: *Aspergillus flavus*; Afu: *Aspergillus fumigates*; Ala: *Alternaria alternata*; Al: *Alternaria* spp.; R: *Rhizopus* spp.; C: *Curvularia* spp.; P: *Penicillium* spp.; H: *Helminthosporium* spp.; T: *Trichoderma* spp.; F: *Fusarium* spp.

objects by impaction which causes severe losses to valuable property. In the new millennium, when we are concerned with the conservation of these valuable material in library detailed studies are required to know the behavior of these biodeteriogens like fungi and to find out methods to control them.

The awareness of microbial deterioration of library materials came only after nineteen century (Orlita, 1977; Gallo, 1993). The first scientific study, which discusses microbial deterioration of papers, comes from France, 1917 (See, 1919). The history of insect cause deterioration of library materials exceeds 1000 years, while knowledge of microbial deterioration of library materials is only about 90 years old (Zyska, 1993). *Aspergillus* and *Penicillium* are the most frequently isolated fungi. These genera cause the decay of stored products in libraries. In this study we intended to survey the extent of fungal contamination of library materials as one of the main external

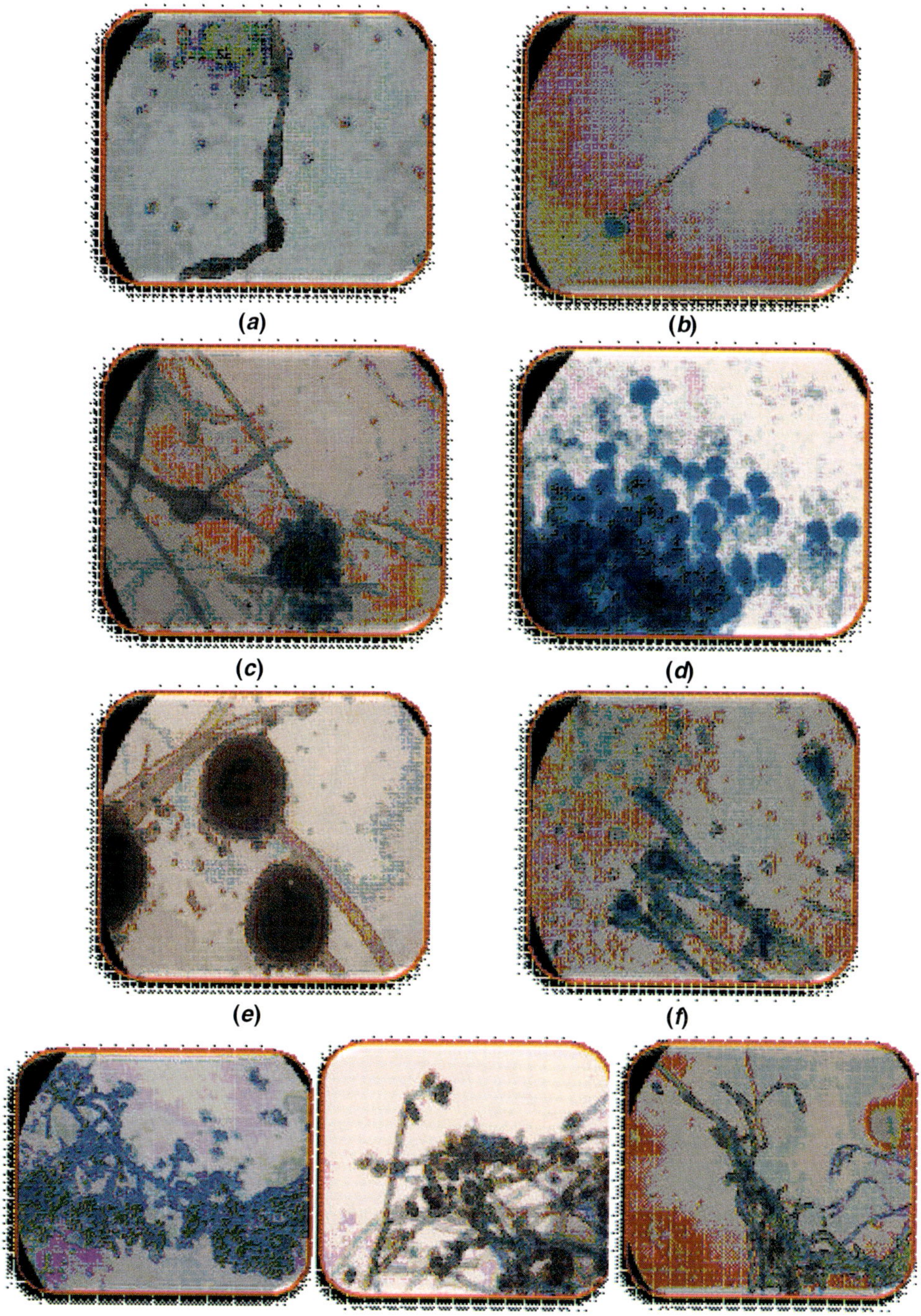

(*a*) (*b*) (*c*) (*d*) (*e*) (*f*) (*g*) (*h*) (*i*)

Figure 26.2: Photos Showing a: *Alternaria alternata*; b: *Aspergillus* spp.; c: *Rhizopus* spp.; d: *Aspergillus fumigatus*; e: *Aspergillus niger*; f: *Penicillium* spp.; g: *Trichoderma* spp.; h: *Curvularia* spp.; i: *Helminthosporium* spp.

factors that influences library material. The libraries under study are from the Wardha city which has a tropical climate. One of the basic aspect lacking is the monitoring and maintenance of temperature and humidity between 18 to 22°C. and 55 per cent respectively. These conditions make the libraries very vulnerable to fungi and fungal spores which can thrive well. Since in our study of five libraries also show a high prevalence of *Aspergillus, Alternaria, Curvularia* and *Penicillium* their presence as contamination agents on studied manuscripts, are not un-expected.

Among all different methods used for fungal isolation, maintained culture of isolated fungi was most sensitive than direct microscopic examination and also was more reliable than macroscopic inspection. So it could be recommended that culture should be used for routine inspections especially in libraries containing old and valuable archival materials. Bearing in mind that fungi have been isolated from a great number of books, activity of fungi and deterioration of manuscripts are a real hazard to library materials. For prevention of damage and preserving library collections, environmental conditions should be adjusted in a way that fungal growth dimishes (Florian, 1994). The optimum temperature for this purpose should be between 18 to 22°C. and humidity should be adjusted below 55 per cent (Florian, 1994). Therefore, sanitary measures should be considered during the construction of library buildings in order to prevent exposure to fungi and to minimize fungal growth (Florian, 1994). Along with these activities, various methods of disinfection such as chemical methods and radiation should be considered for elimination of contaminating fungi (Ballard and Baer, 1986; Haines and Kohler, 1986; Adamo *et al.*, 2001).

Acknowledgements

We wish to thank the Principals of the Colleges for giving permission to conduct air sampling in their respective libraries and we are also thankful to the Head of the Department of Botany, Hislop College, Nagpur for laboratory facilities.

References

Adamo, M., Brizzi, M., Magaudda, G. and Martinelli, G. (2001). Gamma radiation treatment of paper in different environmental conditions: chemical, physical and microbiological analysis. *Restaurator*, **22**: 107-31.

Babu, M. (1983). Aerobiological studies at Aurangabad. *Ph.D. Thesis,* Marathwada University, Aurangabad.

Ballard, M.W. and Baer, N.S. (1986). Ethylene oxide fumigation: Results and risk assessment. *Restaurator*, **7**: 143-68.

Florian, M.L.E. (1994). Conidial fungi (mould, mildew) biology: A basis for logical prevention, eradication and treatment for museum and archival collections, *Leather Conservation News*, **10**: 1-28.

Gallo, F. (1993). Aerobiological research and problems in libraries. *Aerobiologia*, **9**: 117-30.

Haines, J.H. and Stuart, A.K. (1986). An evaluation of orthophenyl phenol as a fungicidal fumigant for archives and libraries. *J. American Inst. Conserv.*, **25**: 49-55

Orlita, A. (1977). The occurrence of fungi on book leather bindings from the Baroque period. *Int. Bio-deterioration Bulletin*, **13**: 45-7

Patil Mukundraj, B., Kamble, S. A. and Pillai, Shanta G. (2009). Airborne bioparticles in the industrial area. *Bionanofrontire*, Science Day Special issue, p. 85-86.

Pillai, Shanta G. and Patil, Mukundraj B. (2007). Effect of threshing on air pollution of indoor environment. *Ind. J. Aerobiol.*, **20**: 63-65.

Sinha, A., Singh, M.K. and Kumar, R. (1998). Aerofungi: An important atmospheric biopollutant at atmosphere. *Ind. J. Aerobiol.*, **11**: 19-23.

Sreeramulu, T. and Ramalingam, A. (1967). A two year study of the airspora of Paddy fields near Visakhapatanam. *Ind. J. Agric. Sci.*, **36**: 112-132.

Tilak, S.T. and Vishwe, D.B. (1975). Microbial content of air inside library. *Biovigyanam*, **1**: 187-190.

Tilak, S. T. and Chakre, O.J. (1978). Atmospheric concentration of *Cleviceps fusiforms* over Bajra fields in relation to their environment factors. *IV Int. Conf. Palynol.*, 1976-77, Lucknow, Abst. p. 181-182.

Tilak, S.T. and Pillai, S.G. (1988). Fungi in library: An aerobiological survey. *Ind. J. Aerobiol.*, **1**(2): 92-94.

Tiwari, K.L. (2005). Studies on aeromycoflora of dairy area at Raipur (C.G.), India. *Flora and Fauna* **2** (2): 195-196.

Verma, K.S. and Srivastava (2004). Airborne fungi of poultry environment. *Advances in Pollen Spore Research*, **22**: 93-94.

Wessel, C.J. (1970). Environmental factors affecting the permanence of library materials. In: *Deterioration and Preservation of Library Materials*. The University of Chicago Press, Chicago, pp. 39-84.

Zyska, B. (1993). *Preservation of Library Materials, Vol.* 2. Factor Deteriorating Materials in Library collections. Uniwersytet Slgski, Katowice. Cited in Zyska browinslaw, (1997), Fungi isolated from library materials: A review of the literature. *Int. Bio-deterioration and Bio-degradation*, **40**: 43-51

Chapter 27

Effect of Basidiocarp Extract of Different Mushrooms on Growth of Mould Fungi

D.K. Kirwale[1], S.S. Bharade[2] and D.S. Mukadam[3]*

ABSTRACT

The fleshy fruiting bodies of different mushrooms are known to contain poisonous or non poisonous components which may affect growth in microorganisms either stimulatory or inhibitory way. In order to study this effect the fruiting bodies of selected 10 fleshy fungi from this region were used for the study. The water extract of these fruiting bodies were incorporated separately in Agar media (PDA). The PDA without extract served as a control. On such extract containing agar media growth of 9 test fungi was studied. It is clear from the results that growth of *Aspergillus flavus* was stimulated in the medium containing extract of *Pleurotus sajor-caju, Agrocybe, Volvariella, Coprinus* and *Polyporus* species. Similarly, *A. niger* showed stimulation of growth in the presence of *Pleurotus, Agaricus* and *Mucidula* while it was inhibited in the extract of other fleshy fungi.

---

1 Department of Botany, Mahatma Phule Arts, Science and Commerce College, Panvel, M.S., India

2 Department of Botany, Badrinarayan Barwale College, Jalna, M.S., India

3 Department of Botany, Dr.Babasaheb Ambedkar Marathwada University, Aurangabad, M.S., India

* Corresponding Author E-mail: devidaskirwale@yahoo.com

The two species of *Alternaria* were stimulated in the presence of *Pleurotus* and *Agaricus* while inhibited in the extract of other mushrooms except *Mucidula* and *Polyporus* for *Alternaria parii.*

Keywords: *Basidiocarp, Fruiting bodies, Mould fungi, Mushroom.*

Introduction

Mushrooms provide a rich addition to the diet in the form of proteins, carbohydrates, valuable salts and vitamins. As food the nutritional value of mushrooms lies between meat and vegetables L. Investigations by Lintzel (1941) indicate that 100 to 200 g of mushrooms (dry weight) are required to maintain nutritional balance in a normal human being weighing 70kg Zakia and Rajrathnam (1982).. They equated the nutritive value of mushrooms to that of muscles protein (Li and Yang, 1989). Experiments proved that mushrooms are well suited to supplement diets Verma *et al.* (1987). The aim of the present study was to examine the effect of extracts of fruiting bodies of selected 10 fleshy fungi (Mushrooms) are known to contain poisonous or no-poisonous components which may affect growth of mould fungi. Vijay *et al.* (1987).

Material and Methods

The fleshy fruiting bodies of different mushrooms are known to contain poisonous or non-poisonous components which may affect growth in microorganisms either stimulatory or inhibitory way Sinden (1972). In order to study this effect the fruiting bodies of selected 10 fleshy fungi from Marathwada region were used for the study. The water extracts of these fruiting bodies were incorporated separately in PDA agar media (Rai and Sohi, 1998). The PDA without extract served the control on such extract containing agar media growth of 9 test fungi (saprophytes and plant pathogenic) was studied and recorded the observations.

Results and Discussion

It is clear from the Table 27.1 that growth of *Aspergillus flavus* was stimulated in the medium containing extract of *Pleurotus Sajor-Caju, Agrocybe, Volvariella, Coprinus* and *Polyporus.* Similarly, *A. niger* showed stimulation of growth in the presence of *Pleurotus, Agaricus* and *Mucidula* while it was inhibited in the extract of other fleshy fungi. The two species of *Alternaria* were stimulated in the presence of *Pleurotus* and *Agaricus* while inhibited in the extract of other mushrooms except *Mucidula* and *Polyporus* for *Alternaria porri.* It was interesting to observe that only Lepiota extract inhibited growth of *Penicillium* but remaining all mushrooms stimulated the growth. Similarly, the growth of *Trichoderma viride* was accelerated in the extract of most of the mushrooms and was inhibited to the maximum due to *Volvariella.* The Fungi *Cladosporium allii* and *Macrophomina phaseolina* showed their growth in increased manner slightly over the control due to extracts of all the basidiocarps of mushroom tested.

Table 27.1: Effect of Basidiocarp Extracts of Different Mushrooms on Growth of Mould Fungi

Moulds	*Diameter of Colony in Extracts of Mushrooms (mm)*										
	Control PDA	*1*	*2*	*3*	*4*	*5*	*6*	*7*	*8*	*9*	*10*
Aspergillus flavus	32.4	36.5	8.4	10.2	8.8	35.8	26.7	3.0	12.5	35.0	10.0
Aspergillus niger	33.1	38.8	8.9	19.8	9.5	33.5	35.1	9.5	20.2	24.2	12.5
Alternaria solani	25.2	28.9	7.4	8.8	12.0	30.4	26.4	8.8	21.5	26.3	25.7
Alternaria porri	24.6	30.2	9.1	9.7	8.7	28.9	9.4	11.3	11.2	9.0	13.2
Penicillium oxysporum	20.2	29.1	13.5	28.5	26.6	23.6	22.9	26.4	20.7	27.7	28.9
Helminthosporium spp.	10.7	2.4	15.0	12.6	11.4	12.5	15.5	8.5	10.2	15.6	12.6
Trichoderma viride	35.8	40.3	31.9	25.9	27.0	40.4	41.8	18.6	23.1	46.1	40.7
Cladosporium allii	9.0	12.1	8.8	8.4	9.0	10.9	11.2	11.2	9.5	14.0	9.8
Macrophomina phaseolina	9.6	13.4	14.5	8.8	11.2	11.3	10.6	12.4	9.9	20.4	8.5

1: *Pleurotus sajor caju*; 2: *Lepiota* spp.; 3: *Amanita* spp.; 4: *Agrocybe* spp; 5: *Agaricus* spp.; 6: *Mucidula* spp.; 7: *Volvariella*; 8: *Coprinus*; 9: *Ganoderma*; 10: *Polyporus*.

References

Li, Y.Y. and Yang, L. (1989). Nutritive value of waste from the cultivation of *Pleurotus sapidus* In: *Edible Fungi of China*, **1**: 31.

Lintzel, W. (1941).The nutritional value of edible mushroom protein. *Biochem. Acta*, **308**: 413-419.

Rai, R.D. and Sohi, H.S. (1998). How protein rich are mushrooms? *Indian Hort.*, **33**(2): 2-3.

Sinden, J.W. (1972). Biological control of pathogens and weed moulds in mushroom culture. *Ann. Rev. Phytopath.*, **9**: 411.

Vijay, B. and Sohi, H.S. (1987). Fungal competitors of *Pleurotus sajor caju* (Fr.) Sing. *Mushroom J. for the Tropics*, **9**: 29-35.

Verma, A., Kesherwani, G.P., Sharma, V.K., Keshwal, R.L. and Singh, P. (1987). Nutritional evaluation of dehydrated mushrooms. *Ind. J. Nutr. Dietet.*, **24**: 380-384.

Zakia, B. and Rajrathnam, S. (1982). Effect of protein supplementation on yield of mushroom. *Mush. J.*, **102**: 178-179.

Chapter 28

Evaluation of *In vitro* and *In vivo* Antibacterial Properties of *Nothapodytes nimmoniana* (Grah.) Mabb. and Standardization of its Micropropagation Technique

Pallavi P. Borate[1] *and Siddheshwar D. Disale*[1]

ABSTRACT

Nothapodytes nimmoniana (Grah.) Mabb. (Lcacinaceae) is a medium sized woody tree occurring moderately in almost entire Sahyadri ranges (western ghats of Maharashtra, India). It has gained international importance due to its recently identified pharmacological and curative properties. Due to which it is being cut and smuggled by poachers. This is tremendously reducing its population and becoming vulnerable species. In the present study, attempt was made to

---

1 Department of Biotechnology, D.B.J. College, Chiplun – 415 605, M.S., India

evaluate the antibacterial activity of leaf and callus extracts and to standardize the process of micro-propagation from embryo culture by using MS medium and plant growth regulators like picloram, 2,4-D, AB, NAA and 2,4,5-T. The extract showed antibacterial activity against five different organisms *viz. Escherichia coli, Staphylococcus aureus, Pseudomonas aerogenosa, Bacillus subtilis* and *Bacillus cereus.* Micropropagation studies showed better growth in MS medium supplemented with plant growth regulator, Picloram for all concentrations studied. In case of MS supplemented with 2, 4–D, explants showed poor response. However, the response is maximum if MS is supplemented with combination of 2, 4–D and AB followed by NAA and 2, 4, 5–T. The cultivable plantlets can be produced in approximately three months. Under licenced conditions, if, the farmers are promoted for the cultivation of this plant, the plantlets are provided to the farmers in subsidized rates and the crop is purchased by government agency, it will definitely not only help to propagate the species and harvest for medicinal purpose but also help to earn and uplift the status of farmers in the hilly area.

Keywords: Anticancer property, Camptothecin, Growth supplements, Micropropagation, Nothapodytes nimmoniana.

Introduction

Nothapodytes nimmoniana (Grah.) Mabb. (Icacinaceae) is a medium sized woody tree occurring moderately in almost entire Sahyadri ranges (Western Ghats of Maharashtra, India). It has gained international importance due to its newly identified pharmacological and curative properties. Due to which it is being cut and smuggled by poachers. This is tremendously reducing its population and becoming vulnerable species.

It is a rich source of potent alkaloid camptothecin (CPT) and 9-methoxy-camptothecin (Govindachari *et al.*, 1972 and Fulzele *et al.*, 2001). Biological screening revealed that camptothecin and its derivatives have anti-cancer activity. Direct camptothecin is cytotoxic, but its derivatives are most effective in the treatment of cancer and are being used throughout the world. Recently two semi-synthetic analogues *viz.* irrinotecan (CPT-11) and topotecan have been of major interest in the present decade and approved for the treatment of cancer (Kingsbury *et al.*, 1991 and Sawadac *et al.*, 1991). Production of secondary metabolites through tissue culture and their evaluation seems to be the only sustainable method for their production (Fulzele *et al.*, 2002).

In present study, attempts are made to standardize the procedure for its micropropagation through embryo culture and produce its plantlets that can; not only help conserve the species but also propagate the species. Along with this, antimicrobial activity from leaves of *Nothopodytes nimmoniana* (*in vivo*) and cells obtained from *in vitro* grown cultures were investigated.

Materials and Methods

The mature fruits (berries) of *Nothapodytes nimmoniana* were obtained from natural habitat. They were thoroughly washed and soaked in tap water for 8 to 10 hours.

Then rinsed with distilled water and operated in LAF (Laminar Air Flow) to collect the seeds. The excised seeds were sterilized by using 0.1 per cent $HgCl_2$ for 7 to 8 minutes. Aseptic conditions are strictly followed. The seeds were dissected to obtain embryos. These embryos are treated as explants. For the growth of explants, plain MS medium (Murashige T. and Skoog, F. A, 1962) and MS medium supplemented with different concentrations of auxins such as 2, 4-dichlorophenoxy acetic acid (2, 4-D), Naphthalene acetic acid (NAA) and Picloram was used. The cultures were incubated at 25°C (± 5 °C) and 2000 Lux illumination for 8 to 16 hours.

Organic solvents like acetone and methanol were used for extraction of active components from callus cells and fresh leaves of this plant. The Soxhlet extraction procedure was used (Furniss 1994). It involves following steps.

1. 30 gm of (dried leaves and fresh callus tissue) plant material is suspended in 250 ml appropriate organic solvent.
2. The extract was filtered and distilled off by using rotary evaporator at 70°C and 60oC respectively for acetone and methanol respectively to furnish the desired brownish green residue.
3. The yield of this process is 3.900 gm.
4. The purified residues were then re-suspended in acetone and methanol at 1, 3, 5 and 10 per cent concentrations.

These concentrations were tested for antimicrobial activities by using six different organisms *viz. Escherechia coli, Staphylococcus aureus, Pseudomonas aerogenosa, Bacillus subtilis* and *Bacillus cereus.*The antimicrobial activity was tested by using agar cup bioassay method (Shimpi *et al.,* 2002). Nutrient agar and Saboraud's agar medium were used for testing antimicrobial activity.

The detailed day wise observations of embryo cultures in different concentrations of auxins are tabulated in Tables 28.1–28.3. The cultured embryos and plantlets were presented in Plate 28.1. The observations of Agar cup bio-assay are tabulated in Tables 28.4–28.9 and antimicrobial activities are presented in Plate 28.2.

Table 28.1: Growth Performances of Excised Embryos of *N. nimmoniana* on Different Concentrations of Picloram

Conc. of Picloram	*Date of Inoculation*	*After 7 Days*	*After 14 Days*	*After 21 Days*	*Response*
1mg/lit	5/12/2007	Green colour and swelling	Green colour Curling of explant Initiation of callus	Greenish yellow callus growth	+ + + +
2mg/lit	5/12/2007	Green colour and swelling	Initiation of callus at the point of radical	Yellow callus growth	+ + +
3mg/lit	5/12/2007	Green colour and swelling	Initiation of callus at the point of radical	Greenish yellow callus growth	+ + +

Table 28.2: Growth Performances of Excised Embryos of *N. nimmoniana* on Different Concentrations of NAA

Conc. of Picloram	*Date of Inoculation*	*After 7 Days*	*After 14 Days*	*After 21 Days*	*Response*
1mg/lit	9/12/2007	Green colour and swelling	Green colour Initiation of callus	Greenish yellow coloured callus growth	+
2mg/lit	9/12/2007	Green colour and swelling	Pale yellow colour Initiation of callus	Yellow coloured callus growth	+
3mg/lit	9/12/2007	Green colour and swelling	Dark green colour Initiation of callus	Curling of explant Dark green coloured callus growth	+ + +
4mg/lit	9/12/2007	Green colour and swelling	Dark green colour Initiation of callus	Curling of explant Greenish yellow coloured callus growth	+ + + +

Table 28.3: Growth Performances of Excised Embryos of *N. nimmoniana* on Different Concentrations of 2, 4-D

Conc. of Picloram	*Date of Inoculation*	*After 7 Days*	*After 14 Days*	*After 21 Days*	*Response*
1mg/lit	13/12/2007	Green colour and swelling	Pale yellow coloured swelling	Yellow coloured callus growth	+
2mg/lit	13/12/2007	Green colour and swelling	Yellow colour Initiation of callus	Yellow coloured callus growth	+
3mg/lit	13/12/2007	Green colour and swelling	Yellow colour Initiation of callus	Yellow coloured callus growth	+ + +
4mg/lit	13/12/2007	Green colour and swelling	Yellow colour Initiation of callus	Yellow coloured callus growth	+ + + +

Table 29.4: Response of Organisms to the Leaf Extract

Name of Organism	*Zone of Inhibition (Conc.) (in mm)*				*Control*
	1 per cent	*3 per cent*	*5 per cent*	*10 per cent*	
Escherechia coli	2	4	3	3	No growth
Staphylococcus aureus	1	3	4	3	No growth
Pseudomonas aerogenosa	2	3	1	2	No growth
Bacillus subtilis	2	3	4	9	No growth
Bacillus cereus	3	5	6	4	No growth

Table 29.5: Response of Organisms to the Extract Obtained from Callus Grown on MS + Picloram (1mg/lit)

Name of Organism	*Zone of Inhibition (Conc.) (in mm)*				*Control*
	1 per cent	*3 per cent*	*5 per cent*	*10 per cent*	
Escherechia coli	4	–	–	–	No growth
Staphylococcus aureus	1	2	–	–	No growth
Pseudomonas aerogenosa	3	2	1	1	No growth
Bacillus subtilis	4	2	–	–	No growth
Bacillus cereus	3	2	1	1	No growth

Table 29.6: Response of Organisms to the Extract Obtained from Callus Grown on MS + Picloram (2mg/lit)

Name of Organism	*Zone of Inhibition (Conc.) (in mm)*				*Control*
	1 per cent	*3 per cent*	*5 per cent*	*10 per cent*	
Escherechia coli	–	–	–	–	No growth
Staphylococcus aureus	2	2	1	2	No growth
Pseudomonas aerogenosa	2	–	–	–	No growth
Bacillus subtilis	2	1	2	4	No growth
Bacillus cereus	5	2	1	1	No growth

Table 29.7: Response of Organisms to the Extract Obtained from Callus Grown on MS + Picloram (3mg/lit)

Name of Organism	*Zone of Inhibition (Conc.) (in mm)*				*Control*
	1 per cent	*3 per cent*	*5 per cent*	*10 per cent*	
Escherechia coli	–	–	–	–	No growth
Staphylococcus aureus	2	6	3	1	No growth
Pseudomonas aerogenosa	–	–	–	–	No growth
Bacillus subtilis	–	3	–	4	No growth
Bacillus cereus	–	–	–	1	No growth

Note: '–' (hyphen) indicates no zone of inhibition.

Result and Discussion

Embryo Cultures

The tabulated observations for Picloram indicate that the explants showed very good (+ + + +) and good (+ + +) responses after 21 days. The growth of callus shows very good responses with greenish yellow colour. In case of NAA, explants showed very good (+ + + +) response with greenish yellow colour at 4mg/lit concentration as

Plate 28.1: Cultured Embryos and Plantlets of *N. nimmoniana*

Control: No zone of inhibition.

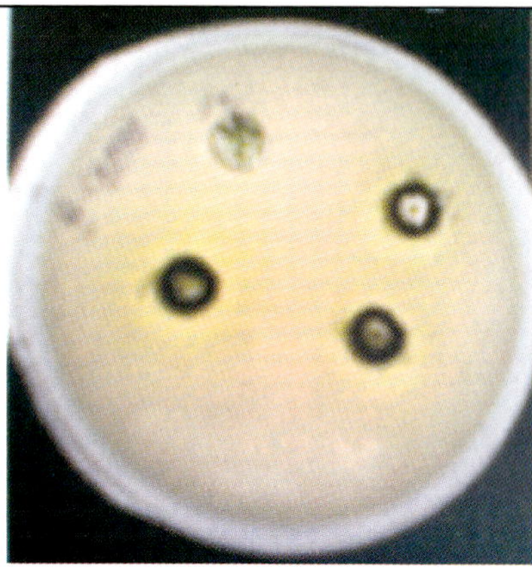

Test of extract of *in vivo* plant material grown on MS + Picloram 2 mg/l on *B. cereus*

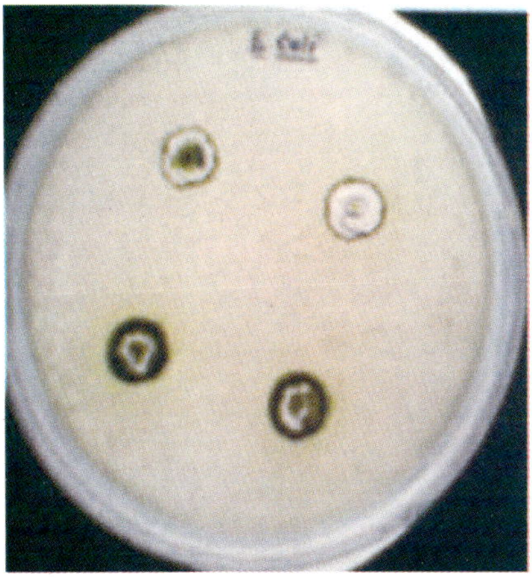

Zone of inhibition for *E. coli* against extractobtained from *in vivo* plant material.

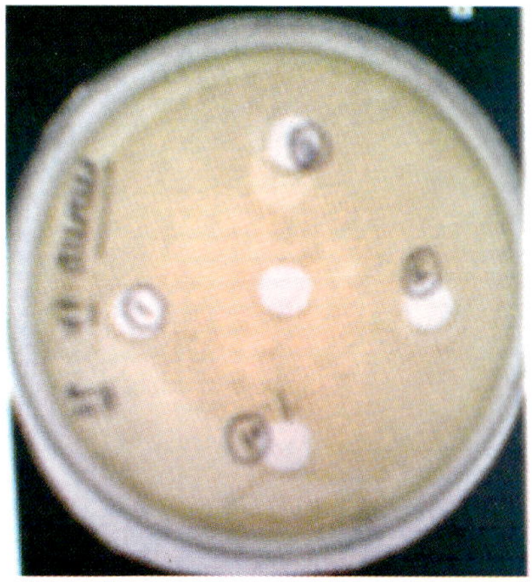

Zone of inhibition for *S. aureus* against extract obtained from *in vivo* plant material grown on MS + picloram 3 mg/l

Plate 28.2: Antibacterial activities of extracts of *N. nimmoniana*

compare to 3mg/lit concentration, at which explant showed good response. The response decreased as concentration of NAA decreased. Thengane and his colleagues also have reported similar correlation (Thengane *et al.*, 2003). In present study, better picture is observed with plant growth regulator, Picloram for all concentrations studied. In case of MS supplemented with different concentrations of 2, 4–D showed poor response. It has been shown that response of explants is maximum if MS is supplemented with combination of 2, 4–D and AB followed by NAA and 2, 4, 5–T. hence 2, 4–D alone is not sufficient to induce callus formation in present plant.

Table 29.8: Response of Organisms to the Extract Obtained from Callus Grown on MS + NAA (3mg/lit)

Name of Organism	*Zone of Inhibition (Conc.) (in mm)*				*Control*
	1 per cent	*3 per cent*	*5 per cent*	*10 per cent*	
Escherechia coli	1	3	–	–	No growth
Staphylococcus aureus	–	–	–	–	No growth
Pseudomonas aerogenosa	1	–	–	–	No growth
Bacillus subtilis	–	–	–	–	No growth
Bacillus cereus	1	2	1	1	No growth

Table 29.9: Response of Organisms to the Extract Obtained from Callus Grown on MS + NAA (4mg/lit)

Name of Organism	*Zone of Inhibition (Conc.) (in mm)*				*Control*
	1 per cent	*3 per cent*	*5 per cent*	*10 per cent*	
Escherechia coli	3	–	–	–	No growth
Staphylococcus aureus	–	2	–	–	No growth
Pseudomonas aerogenosa	2	–	–	–	No growth
Bacillus subtilis	–	–	–	–	No growth
Bacillus cereus	2	3	3	4	No growth

Note: '–' (hyphen) indicates no zone of inhibition.

Antimicrobial Activity

In case of antimicrobial tests, all controls showed no activity. The zone of inhibition for each extract and solvent is found to be different. However increasing concentrations of extracts induced more zone of inhibition. In case of leaf extracts, mostly mixed results are obtained. In case of acetone extracts, no zone of inhibition was observed in case of *E. coli, P. aerogenosa, B. cereus, B. subtilis* and *St. aureus*. Whereas, in methanol extracts, highest activity was observed against *B. subtilis, B. cereus, E. coli, S. aureus* and *P. aerogenosa*.

From the above results, it can be concluded that plant extracts *in vivo* and *in vitro* have great potential as antimicrobial agents against microorganisms such as

Escherechia coli, Staphylococcus aureus, Pseudomonas aerogenosa, Bacillus subtilis and *Bacillus cereus*. These extracts are found to be very active as antimicrobial agents. Therefore these can be used against infectious diseases caused by these microbes. *N. nimmoniana* shows maximum antibacterial activity. Therefore, it can be used to discover bioactive natural compounds that may serve for development of new pharmaceuticals that address heither to unmate therapeutic needs. Such screening of various natural organic compounds and identifying active agents is the need of the hour, because successful prediction of lead molecule and biomedicinal properties at the onset of drug discovery will pay off later in drug development. Besides, tissue culture proves to be boon for sustainable exploitation of the plant resources.

Acknowledgement

Authors are thankful to the Department of Biotechnology, Shivaji University Kolhapur and D. B. J. College, Chiplun for kind co-operation in conducting the research work in their laboratories.

References

Fulzele, D.P., Satdive, R.K. and Pol, B.B. (2002). Untransformed root cultures of Nothapodytes foetida and production of camptothecin. *Plant Cell, Tissue and Organ Culture*, **69**: 285–288.

Fulzele, D.P., Satdive, R.K. and Pol, B.B. (2001). Growth and production of camptothecin by cell suspension cultures of *Nothapodytes foetida*. *Planta Med.*, **67**: 150–152.

Furniss, B.S. *et al.* (1994). *Vogel's Text Book of Practical Organic Chemistry*, 5th Edn. ELBS and Longman, London.

Govindachari, T.R. and Vishwanathan (1972). Alkaloids of *Mappia foetida*. *Phytochemistry*, **11**: 3529-3531.

Kingsbury, W.D., Boehm, J.C. and Jakas, D.R. (1991). Synthesis of water-soluble (aminoalkyl) camptothecin analogs: inhibition of topoisomerase I and antitumor activity. *J. Amer. Chem. Soc.*, **34**: 98–107.

Murashige, T. and Skoog, F.A. (1962). Revised medium for rapid growth and bio-assay with tobacco tissue cultures. *Physiologia plantarum*, **15**: 473–497.

Sawada, S., Okajima, S., Aiyama, R., Nokata, K., Furuta, T., Yokokura, T., Sugino, E., Yamaguchi, K. and Miyasaka, T. (1991). Synthesis and antitumor activity of 20(s)-camptothecin derivatives: Carbamate linked water soluble derivatives of 7-ethyl-10-hydroxycamptothecin. *Chem. Pharm. Bull.*, **39**: 1446–1450.

Shimpi, S.R., Chaudhari, L.S., Barambe, S.M. and Kharche, A.T. (2002). Evaluation of antimicrobial activity of organic extract of leaves of *Aristolochia bracteata*. *J. Pest. Res.*, **17(1)**: 16-18.

Thengane, S.R., Kulkarni, D.K., Shrikhande, V.A., Joshi, S.P., Sonawane, K.B. and Krishnamurthy, K.V. (2003). Influence of medium composition on callus induction and camptothecin(s) accumulation in *Nothapodytes foetida*. *Plant Cell, Tissue and Organ Culture*, **72**: 247–251.

Chapter 29

Induced Variation in Plant Height Due to Physical (Gamma Rays) and Chemical (EMS) Mutagenic Treatmens in Groundnut (*Arachis hypogaea* L.)

R.A. Satpute[1]* and S.W. Suradkar[2]

ABSTRACT

Groundnut (*Arachis hypogaea* L.) is a member of family Fabaceae. It is economically important oilseed and food crop. The present investigation deals with the induced variation in plant height of two varieties of groundnut (*A. hypogaea* L.) *i.e.* TAG-24 and AK-159 by the treatment of gamma rays (10kR, 15kR and 20kR) and EMS (0.05 per cent, 0.10 per cent and 0.15 per cent) in three different doses and concentration respectively. About 22 Tall and 19 Dwarf mutants were found in M3 generation of variety AK-159, the plant height was increased by 53.84 per cent and decreased by 28.20 per cent respectively. 18 Tall

1 Department of Botany, Government Institute of Science and Research Center, Aurangabad – 431 004, M.S., India

* Corresponding Author E-mail: santoshbot214@gmil.com

and 11 Dwarf mutant were recorded form variety TAG-24. In the same variety plant height was increased by 21.21 per cent and decreased by 24.24 per cent But from the whole population it is concluded that Chemical mutagen *i.e.* EMS concentration 0.10 per cent is more effective in increasing plant height in both varieties and 20kR dose of gamma rays in variety AK-159 and 10kR of gamma rays in Variety TAG-24.

Keywords: *Groundnut, Mutation and Plant height.*

Introduction

Groundnut is an important monoecious annual legume in the world mainly grown for oilseed, food and animal feed (Pande *et al.*, 2003; Upadhyaya *et al.*, 2006). It contains oil (44-56 per cent), Protein (22-30 per cent) and it is also a good source of minerals (Phosphorus, calcium, magnesium and potassium) and vitamins (E, K, and B group) and phytosterols to increase consumer preference after value addition (Savage and Keenan, 1940).

Due to lack of improved varieties and seed availability, farmers recycle seeds which further complicates the situation (Doss *et al.*, 2003). This has raised concerns to breeders, farmers and policy makers on the breeding for better groundnut varieties and their subsequent introduction in the country. Mutation breeding as a source of increasing variability and could confer specific improvement without significantly altering its acceptable phenotype. (Ojomo *et al.*, 1979). Wide spectrum of genetic variability has been induced in groundnut using both physical and chemical mutagens in order to utilize it in groundnut improvement and inheritance studies (Ashri, 1970, Gowada *et al.*, 1996). Earlier reports in groundnut showed that several mutations affected qualitative and quantitative traits such as leaf size, shape and colour, plant height, plant habit, flower colour, pod and seed traits (Dwivedi *et al.*, 1996, Desale *et al.*, 1986).Groundnut dwarf mutants were induced using X rays (Patil., 1966), gamma rays (Patil and Mouli, 1984), Ethyl methane sulfonate (Gowda *et al.*, 1996).

The present investigation deals with the induced variation in plant height of two varieties of groundnut (*Arachis hypogaea* L.) *i.e.* TAG-24 and AK-159 by the treatment of gamma rays (10kR, 15kR and 20kR) and EMS (0.05 per cent, 0.10 per cent and 0.15 per cent) in three different doses and concentration respectively.

Material and Methods

The seed sample was obtained from the Department of Crop Research Unit (Oilseeds), Dr. Panjabrao Deshmukh Krishi Vidhya peeth, Akola-444 104. (M.S.) was used in present study. Healthy and dry seeds of groundnut having uniform size and equilibrated to moisture level of 7 per cent were packed in small polyethene bags and irradiated to Co^{60} at three different doses *viz.* 10 kR, 15 kR and 20 kR in the gamma chamber of Department of Biophysics, Government Institute of Science, Aurangabad. (M.S.) India.

Irradiated seeds of each treatment were sown in field for raising M_1 generation following randomized block design (RBD) with three replicate of each dose and variety, consisting 100 seeds of each along with control in the field. While sowing the seeds in field the spacing between plants was 15 cm and rows 35 cm were maintained. The experimental area for M_2 and M_3 and spacing are same as those used in M_1 generations. Critical screening was done though the M_1, M_2 and M_3 generation for plant height at maturity.

Result and Discussion

The induced mutagenesis of both the varieties of groundnut shows several mutants affecting plant height in M_2 and M_3 generation. 22 Tall and 19 Dwarf mutants were found in M3 generation of variety AK-159, the plant height was increased by 53.84 per cent and decreased by 28.20 per cent respectively. 18 Tall and 11 Dwarf mutants were recorded form variety TAG-24. In the same variety plant height was increased by 21.21 per cent and decreased by 24.24 per cent. Data regarding plant height are presented in Tables 29.1 and 29.2.

Table 29.1: Effect of Mutagens on Plant Height in M_3 Generation of *Arachis hypogaea* (L.) Variety-TAG-24

Mutagen	*Concentration*	*Plant Height at Maturity (cm)*	*S.E.±*
Control	Control	19.8	±0.9539
EMS (per cent)	0.05 per cent	23.166	±0.5811
	0.10 per cent	23.166	±3.5044
	0.15 per cent	18.266	±1.5898
Gamma ray (kR)	10 kR	18.366	±2.3397
	15 kR	20.2333	±1.8746
	20 kR	22.2	±2.2912

S.E. ± = Standard Error

Table 29.2: Effect of Mutagens on Plant Height in M_3 Generation of *Arachis hypogaea* (L.) Variety-AK-159

Mutagen	*Concentration*	*Plant Height at Maturity (cm)*	*S.E.±*
Control	Control	39	±0.6506
EMS (per cent)	0.05 per cent	38.7	±2.0074
	0.10 per cent	40.0333	±0.7423
	0.15 per cent	36	±2.7392
Gamma ray (kR)	10 kR	37.4333	±3.4531
	15 kR	35.0666	±3.6843
	20 kR	34.4333	±2.3666

S.E. ± = Standard Error

From the whole population it is concluded that Chemical mutagen *i.e.* EMS concentration 0.10 per cent in both varieties, 20kR dose of gamma rays in variety TAG-24 and 10kR dose of gamma rays in Variety AK-159 are more effective in increasing plant height. For Dwarfism EMS 0.15 per cent concentration in both varieties, 10 kR dose of gamma rays in variety TAG-24 and 20 kR dose of garnma rays in variety AK-159 are responsible.

Conclusion

Reduced height in plants shows increase in grain yield was recorded by White (2009) in Barly. It was concluded that dwarf mutation was due to monogenic incomplete dominance. The dwarfism is always desirable change in plant breeding because dwarf variety is compact or dwarf crop plants have many advantages in agriculture, including high yield, denser growth, increased resistance to storm damage, and reduced loss during harvesting.

References

Ashri, A. (1970). A dominant mutation with variable penetrance and expressivity induced by diethylsulfate in peanuts, *Arachis hypogaea* L. *Mutation Res.*, **9**: 473-480.

Desale, S.C., Bhapkar, D.G. and Thombre, M.V. (1986). Inheritance of faint orange flower colour in groundnut. *J. Oilseeds Res.*, **3**: 135-136

Doss, C.R., Mwangi, W., Verkuij, H., De-Groote, H. (2003). Adoption of maize and wheat technologies in eastern Africa: A synthesis of the findings of 22 case studies: *CIMMYT Economics Working Paper* 03-06.Mexico, D.F.: CIMMYT.

Dwivedi, S.L., Singh, A.K. and Nigam, S.N. (1996). Unstable white flower color in groundnut (*Arachis hypogaea* L.). *J. Hered.*, **87**: 247248

Gowda, M.V.C., Nadaf, H.L. and Sheshagiri, R. (1996). The role of mutation in intraspecific differentiation of groundnut (*Arachis hypogaea* L.). *Euphytica*, **90**: 105-113.

Ojomo, A.O., Omueti O, Raji, J.A. and Omueti, O. (1979). Studies in induced mutation in cowpea, 5. The variation in protein content following ionizing radiation, *Nig. J. Appl. Sci.*, **21**: 61-64.

Pande, S., Bandyopadhyay, R., Blümmel, M., Narayana Rao, J., Thomas, D., Navi, S.S. (2003). Disease management factors influencing yield and quality of sorghum and groundnut crop residues. *Field Crops Res.*, **84(1&2)**: 89-103.

Patil, S.H. and Mouli, C. (1984). Preferential segregation of two allelic mutations for small leaf character in groundnut. *Theor. Appl. Genet.*, **67**: 327-332

Patil, S.H. (1966). Mutations induced in groundnut by X rays. *Indian J. Genet.*, **26A**: 334-348

Savage, G. P. and Keenan, J. I. (1994). The composition and nutritive value of groundnut kernels. In: *The Groundnut Crop: Scientific Basis for Improvement*, (Ed.) J. Smart. Champman and Hall, London, pp. 173-213.

Upadhyaya, H.D., Reddy, L.J., Gowda, C.L.L. and Singh, S. (2006). Identification of diverse groundnut germplasm: Sources of early maturity in a core collection. *Field Crops Res.*, **97(2-3)**: 261-271.

White, P.J. (2009). *Induced Mutations Affecting Root Architecture and Mineral Acquisition in Barley*. Food and Agriculture Organization of the United Nations, Rome, pp. 338-340.

Chapter 30

Standardization of Inoculation Methods for Mass Screening of Pigeonpea Wilt

D.N. Dhutraj[1], P.R. Bhandarge[1], M.G. Korde[1] and A.L. Harde[1]

ABSTRACT

Pigeonpea wilt was caused by *Fusarium udum* is a major economically important disease causing heavy yield losses. To know the effective method of inoculation a pot experiment was conducted at College of Agriculture, Latur in Kharif 2009 on susceptible Cv. ICP-2376 under artificial epiphytotic conditions. Methods of inoculation exerted significant influence on the development of Pigeonpea wilt caused by *F. udum*. Maximum disease incidence was recorded when Pigeonpea seeds (Cv. ICP-2376) were sown in the soil inoculated with mass inoculum of *F. udum*. Direct sowing of Pigeonpea seeds in artificially inoculated soil (soil inoculum added @ 2.5 g/100 g of soil) was better (120 DAI=100 per cent). Maximum wilting was recorded (92.5 per cent) and (70.00 per cent) at 35 DAI in spore suspension and water culture technique respectively. Spore suspension technique was appeared to be good for creating wilt in Pigeonpea *in vitro*. Seed inoculation method also gave wilt incidence (85.00 per cent) at 150 DAI. For all inoculation methods less disease incidence was observed at initial inoculation stages. Per cent disease incidence increased with increase in days after inoculation

---

1 College of Agriculture, Latur – 413 512, M.S., India

(DAI) period during all the methods under study. As it is evident, soil inoculation method was found far better than all other methods of inoculation *viz.* stem inoculation, water culture technique and spore inoculation techniques.

Keywords: *DAI, Fusarium wilt, Inoculation method, Pigeonpea.*

Introduction

Pigeonpea yield is considerably reduced due to attack of diseases. More than 50 diseases have been reported to affect pigeonpea, but only a few of them are of having economic importance. These include *Fusarium* wilt, sterility mosaic, *Phytophtora* blight, leaf spots etc. The economical important disease is *Fusarium* wilt. *Fusarium* being the soil borne fungus, was described for the first time in India by Butler (1906). He proved the pathogenicity of this fungus and named *Fusarium udum* Butler as a new species. The losses caused by wilt are about 97 thousand tonnes per annum. Wilt disease caused by *F.udum* Butler was attributed as limiting factor in increasing the production potential of the crop in Maharashtra (Anonymus, 2007-2008). The disease occurs in the range of 0.0 to 90 per cent in the farmers field with overall average of 22.5 per cent (Kannaiyan *et al;* 1984).

Pigeonpea wilt was caused by *F. udum* is a major economically important disease causing heavy yield losses. To know the effective method of inoculation a pot experiment was conducted at College of Agriculture, Latur in *Kharif* 2009 on susceptible Cv. ICP-2376 under artificial epiphytotic conditions.

Material and Methods

An attempt was made to test different inoculation methods for their standardization efficacy. These were evaluated against the pathogen *F. udum.*

Seed Inoculation

Mass culture of the test pathogen *Fusarium* sp. were prepared in Potato dextrose broth and inoculum were loaded on seed surface with the help of vacuum pump. Then seeds of pigeonpea Cv. ICP-2376 were sown (10 seeds/pot) and observations was recorded after 30, 60, 90 120 and 150 days after inoculation (DAI) on per cent wilt incidence. Each treatment was replicated thrice. Uninoculated pots served as control.

Soil Inoculation

Mass culture of the test pathogen *Fusarium* sp. were prepared in Potato dextrose broth and added to the pots containing sterilized soil @ 2.5g/100 g of soil. Then seeds of pigeonpea Cv. ICP-2376 were sown (10 seeds/pot) and observations was recorded after 30, 60, 90 120 and 150 days after inoculation (DAI) on per cent wilt incidence. Each treatment was replicated thrice. Uninoculated pots served as control.

Water Culture Technique

Spore suspensions were prepared from 10 days old culture in sterilized distilled water and adjusted to 3-4 x 10^{6}/ml concentration. Big size test tubes (250 x 20 mm) up to ¾ levels were filled with this adjusted spore suspension. 10 days old seedlings of

ICP-2376 raised in sterilized riverbed sand were transferred into test tubes and seedlings were held in straight position by cotton plug. Four replications were maintained and uninoculated tubes with only sterilized distilled water were kept as control. The seedlings were kept on the benches in the net house. Observations on number of wilted plants were recorded from the initiation of wilting and continued up to 35 days.

Spore Suspension Technique

In this method, fifteen days old seedling of wilt susceptible variety ICP-2376 grown in sterilized soil in 10″ diameter plastic pots were used for inoculation. Spore suspention of *F. udum* from 10 days old culture with inoculums concentration 3-4 $x10^6$/ml were prepared. Seedlings were inoculated by pipetting 5 ml of spore suspension pouring around each seedling. Three replications of each isolates were maintained and uninoculated pots were kept as control. These pots were kept in net house. Observations on wilting of seedling were recorded at 7, 14, 21, 28 and 35 days after inoculation.

Results and Discussion

Observation on wilt incidence or seedling mortality under soil inoculation method were recorded from the 15th day after sowing and continued till flowering or progressive developments. Results from Table 30.1 revealed that wilting percentage at 120 and 150 days after sowing was recorded 100 per cent wilt incidence. Wilt incidence or seedling mortality under seed inoculation method were recorded from the 15th day after sowing. The results indicated that the wilting percentage at 120 and 150 DAS was recorded 77.5 and 85.0 per cent respectively.

Observations on wilt incidence in spore suspension method were recorded at 7, 14, 21, 28 and 35 days after inoculation and data are presented in Table 30.2. Maximum wilting percentage was recorded (92.5 per cent) at 35 days after inoculation. Wilting was initiated at 14 days after inoculation. This method appeared to be good for creating wilt in pigeonpea *in vitro*.

Table 30.1: *In vitro* Effect of DAI (Days After Inoculation) on the Wilt Incidence by Soil Inoculation and Seed Inoculation Techniques

Days After Inoculation	*Seed Inoculation Method*	*Soil Inoculation Method*
	Per cent Wilt Incidence	*Per cent Wilt Incidence*
30 DAI	35.00 (20.51)*	40.00 (23.65)*
60 DAI	45.00 (28.50)	82.50 (55.88)
90 DAI	65.00 (40.64)	90.00 (46.79)
120 DAI	77.50 (51.52)	100 (89.98)
150 DAI	85.00 (58.63)	100 (89.98)
Control	00 (0.0)	00 (0.0)
SE ±	1.83	6.20
CD at 5 per cent	5.51	18.67

* Figures in parenthesis are angular transformed values.

Table 30.2: *In vitro* Effect of DAI (Days After Inoculation) on the Wilt Incidence by Water Culture Technique and Spore Suspension Method

Days After Inoculation	*Water Culture Technique*	*Spore Suspension Method*
	Per cent Wilt Incidence	*Per cent Wilt Incidence*
7 DAI	00 (0.0)	0.0 (0.0)
14 DAI	40.00 (23.65)	17.50 (10.08)
21 DAI	60.00 (37.03)	32.50 (18.98)
28 DAI	60.00 (37.03)	65.00 (40.64)
35 DAI	70.00 (44.70)	92.50 (70.60)
Control	00 (0.0)	00 (0.0)
SE ±	1.58	2.60
CD at 5 per cent	4.71	7.83

Figures in parenthesis are angular transformed values.

Under water culture techniques, observations on number of wilted seedlings recorded at 7, 14, 21, 28 and 35 days after inoculation. Initiation of wilting took place at 14 days after inoculation and at 35 days after inoculation, the wilting percentage was observed 70 per cent. The uninoculated seedling did not showed any wilting appearance (Table 30.2).

The inoculation methods like soil inoculation (at 150 DAI), seed inoculation (at 150 DAI) and spore suspension technique (35 DAI) was recorded 100 per cent, 85.00 per cent and 92.50 per cent wilting and seedlings mortality, respectively. So that this methods appared to be good for creating wilt in pigeonpea *in vitro*. However, the maximum wilting could be recorded only upto 70.00 per cent at 35 DAI in water culture technique, which indicated inadequacy of this method for creating wilt infection in pigeonpea. Out of four inoculation techniques evaluated in the present studies *viz.*, soil inoculation, seed inoculation, spore suspension and water culture techniques, maximum wilting was obtained with soil inoculation (100 per cent) and spore suspension technique (92.50 per cent). Thus soil inoculation and spore suspension method was found to be most suitable for creating higher wilt infection due to *F. udum* in pigeonpea.

Both soil inoculation and spore suspension inoculation in pot culture were, thus, found highly effective in creating high disease pressure artificially and could be efficiently utilized in screening for resistant sources/varieties under pot conditions against *F. udum* wilt in pigeonpea. Among these, spore suspension method in pot culture is a new method of inoculation attempted for screening germplasm against *F. udum*. It was found suitable for providing sufficient inoculum potential leading to good disease development required for different studies like host resistance, variability and inheritance of resistance screening under pot culture conditions. Similar results were also reported by Shashi Mishra and Vishwa Dhar (2005) and Rupinder and Pushpinder (2009).

References

Anonymous (2007-2008). Area, production and productivity of pigeonpea. *Annual Report of AICPIP, Kanpur.*

Butler, E.J. (1906). The wilt disease of pigeonpea and pepper. *Agril. J. India*, **1**: 25-36.

Kannaiyan, J., Nene, Y.L., Reddy, M.V., Ryan, I.G. and T.N. Raju (1984). Prevalence of pigeonpea disease and associated crop losses in Asia. *Africa and America Trop. Pest Manag.*, **30**: 62-71.

Mishra, Shashi and Dhar, Vishwa (2005). Efficient method of inoculation by *Fusarium udum*, the incitant of pigeonpea wilt. *Indian Phythopath.*, **58 (3)** 332-334.

Rupinder, S.K. and Singh, Pushpinder P. (2009). Standardization of inoculation method for *F. oxysporum* f. sp. *melonis*. *Indian Phytopath.*, **62(3)**: 314-318.

Chapter 31

Effect of Phosphorus Level and Land Configuration on Growth and Yield of Niger

A.M. Dhange[1], *P.N. Karanjikar*[1] *and S.T. Rathod*[1]

ABSTRACT

A field experiment was conducted at Agronomy Farm, College of Agriculture, Latur during kharif 2006 on clay loam soil which was medium in available nitrogen and organic carbon, slightly alkaline in reaction with high cation exchange capacity. The experiment was laid out in Factorial Randomized Block Design (FRBD) and replicated thrice. Four land configuration treatments *i.e.* without opening furrows (L_1), opening of furrows after two rows (L_2), opening of furrows after four rows (L_3) and opening of furrows after every alternate row (L_4) with combination of four phosphorus levels *i.e.* (P_0) 0 kg (P_1) 20 kg, (P_2) 40 kg and (P_3) 60 kg P_2O_5 ha^{-1} were tested. The land configuration treatment opening of furrow after every alternate rows (L_4) recorded significantly highest growth and all the yield attributes. The land configuration opening of furrow after every alternate row recorded significantly highest grain yield (341.05 kg ha^{-1}) than without opening of furrow (L_1) and opening of furrow after two rows (L_2). However, if was at par with opening of furrow after four rows (L_3). In case of phosphorus levels application of 60 kg P_2O_5 ha^{-1} recorded significantly highest growth as well as yield attributes. It was at par with 40 kg P_2O_5 ha^{-1}. Every increasing the level of phosphorus significantly improved all the yield attributes, seed yield per ha.

Keywords: *Guizotia abyssinica, Land configuration, Phosphorus level.*

---

1 College of Agriculture, Ambajogai, Marathwada Agricultural University, Parbhani, M.S., India

Introduction

Niger (*Guizotia abyssinica* Cass) belongs to the natural order compositae (Asteraceae) and is known by various local names such as Ramtil, Kalatil, Khurasni, Karala etc. Niger is one of the important oilseed crop of India. It is considered a minor oilseed crop. But it is very important in terms of its oil content, quality and potentiality. The good features of this crop are that it gives reasonable seed yield even under poor growing conditions. It can grow very well under rainfed conditions on poor soils of coarse texture, especially on till slopes and in shallow soils of marginal lands.

Low production of niger is attributed to the fact that the crop is grown in dry farming region, where moisture is the most limiting factor and generally farmers do not give the recommended dose of fertilizers for niger crop. It will be worthwhile to explore the possibility of introducing niger as a sole crop in Marathwada region of Maharashtra under rainfed conditions. The information on suitable land configuration treatments and application of suitable phosphorus level of niger under the edapho-ecological conditions of Marathwada is meager. Therefore, the present investigation was carried out to study the impact of land configuration and phosphorus management on niger during kharif season.

Materials and Methods

The experiment was conducted in FRBD design with 3 replication during monsoon season of 2006-2007 at Latur, Maharashtra. The main plot treatments were 4 land configurations *viz.*, without opening of furrow (L_1), opening of furrow after two rows (L_2), opening of furrow after four rows (L_3) and opening of furrow after every alternate row (L_4) and sub plot treatments were 4 phosphorus levels (0, 20, 40 and 60 kg P_2O_5 ha^{-1}). The soil of experimental site was clayey in texture, low in nitrogen (233.24 kg ha^{-1}), medium in phosphorus (12.81 kg ha^{-1}), high in potassium (612.08 kg ha^{-1}) and slightly alkaline in reaction (7.57) crop was sown by dibbling at 30 cm x 10 cm spacing on 1st Aug. 2006. Full dose of N and treatment wise phosphorus was applied in the experimental plot before sowing.

Results and Discussion

Effect of Phosphorus

Every increase in the level of phosphorus resulted in significant influence on the growth (Table 31.1) and yield attributes (Table 31.2). Increase in seed yield was due to augmenting effect of phosphorus application on all yield attributes since a good supply of phosphorus has been associated with increased root growth, hasten plant maturity and quality of seeds. Phosphate compounds have been shown to be essential for photosynthesis, the inter conversions of carbohydrates and related compounds, amino acid metabolism, fat metabolism and sulphur metabolism for oilseed crops. Application of adequate phosphorous results in increased yield and the proportion of oil stored in the seed. Such increase in seed yield of niger was with application of 60 kg P_2O_5 ha^{-1} and it was at par with application of 40 kg P_2O_5 ha^{-1}. Lowest seed yield was recorded with application of 0 kg P_2O_5 ha^{-1}. Similar results were reported by Kachapur *et al.* (1979).

Table 31.1: Mean Height of Plant (cm), Number of Functional Leaves/Plant, Number of Primacy Branches/Plant and Number of Capsules/Plant, Leaf Area/Plant, No. of Flowers/Plant and Total Dry Matter/Plant as Influenced by Different Treatments at Various Growth Stages

Treatments	*Mean Height of Plant (cm)*	*No. of Functional Leaves/ Plant*	*No. of Primary Branches/ Plant*	*No. of Capsules/ Plant*	*Leaf Area/ Plant (dm^2)*	*No. of Flowers/ Plant*	*Total Dry Matter Weight (g)*
			Land configurations (L)				
L_1–Without opening of furrow	68.96	18.60	8.95	32.10	20.58	16.14	21.11
L_2–Opening of furrow after two rows	73.29	25.30	10.31	38.55	21.62	23.91	23.07
L_3–Opening of furrow after four rows	70.68	20.70	8.60	33.54	20.68	17.90	22.14
L_4–Opening of furrow after alternate rows	74.46	26.18	10.56	42.02	22.44	33.30	23.25
SEm ±	1.61	0.71	0.18	0.60	0.22	0.35	0.01
CD (P=0.05)	4.88	2.27	0.75	1.95	0.65	1.21	0.04
			Phosphorus levels (P)				
P_0–(0 kg P_2O_5 ha^{-1})	69.46	16.35	7.38	30.45	18.89	19.52	22.37
P_1–(20 kg P_2O_5 ha^{-1})	71.05	18.18	8.95	32.95	19.16	20.37	22.35
P_2–(40 kg P_2O_5 ha^{-1})	71.43	25.31	9.81	37.36	23.25	22.85	22.42
P_3–(60 kg P_2O_5 ha^{-1})	75.45	26.93	9.89	39.44	23.93	33.01	22.43
SEm ±	1.66	0.76	0.23	0.65	0.22	0.04	0.01
CD (P=0.05)	4.88	2.27	0.75	1.95	0.65	1.21	0.04
			Interaction (L x P)				
S.E. ±	3.37	1.57	0.51	1.03	0.45	0.84	0.03
CD (P=0.05)	N.S.	N.S.	N.S.	N.S.	N.S.	N.S.	N.S.
General mean	71.85	22.19	9.30	35.80	21.31	22.06	22.39

Table 31.2: Mean Number of Grains per Capsules, Mean Number of Grains per Plant, Test Weight (g), Grain Yield and Biological Yield (kg/ha) as Influenced by Different Treatments at Various Growth Stages

Treatments	*No. of Grains/ Capsule*	*No. of Grain/ Plant*	*Test Weight (g)*	*Grain Yield (kg/ha)*	*Biological Yield (kg/ha)*
Land configurations (L)					
L_1–Without opening of furrow	23.50	425.91	2.98	293.36	535.98
L_2–Opening of furrow after two rows	27.31	517.93	3.98	337.71	647.88
L_3- Opening of furrow after four rows	25.10	467.70	3.20	295.81	543.91
L_4–Opening of furrow after alternate rows	28.83	551.41	3.96	341.05	629.97
SEm ±	1.116	40.84	0.05	12.88	19.71
CD (P=0.05)	3.42	116.26	0.34	37.41	57.04
Phosphorus levels (P)					
P_0–(0 kg P_2O_5 ha^{-1})	23.46	458.15	2.82	279.00	553.13
P_1–(20 kg P_2O_5 ha^{-1})	25.58	465.26	3.10	294.83	581.27
P_2–(40 kg P_2O_5 ha^{-1})	26.75	501.35	4.01	336.86	620.56
P_3–(60 kg P_2O_5 ha^{-1})	28.95	538.20	4.07	357.25	602.77
SEm ±	1.166	40.234	0.01	12.93	19.76
CD (P=0.05)	3.42	116.26	0.34	37.41	57.04
Interaction (L x P)					
SEm ±	2.37	80.50	0.241	25.90	39.57
CD (P=0.05)	N.S.	N.S.	N.S.	N.S.	N.S.
General mean	26.18	490.74	3.94	316.98	589.43

Effect of Land Configuration

Treatment (L_4) opening of furrow after every alternate row recorded significantly higher values of all the growth attributes (Table 31.1) seed yield and biological yield kg ha$_{-}^{1}$ followed by opening of furrow after two rows (L_2) and lowest by without opening of furrow (L_1). Seed yield is a function of yield contributing characters. Hence, increase in seed yield of opening of furrow after every alternate row (L_4) and opening of furrow after two rows (L_2) was due to increase in yield attributes compared with other land configuration treatments *i.e.* (L_1) without of opening of furrow and opening of furrow after four rows (L_3). Chaudhari *et al.* (2001), Dangore *et al.* (2001) and Jogadande *et al.* (2003) also reported such differential yield of niger. Opening of furrow after every alternate row (L_4) produced significantly the highest grain yield than without opening of furrow (L_1) land configuration and it was at par with opening of furrow after two rows (L_2).

The present study clearly indicated that opening of furrow after every alternate row (L_4) and opening of furrow after two rows (L_2) significantly influenced growth characters, yield and oil content with the application of 40 kg P_2O_5 ha^{-1} to niger crop.

References

Chaudhari, C.S., Pawar, W.S., Mendhe, S.N., Nikam, R.R. and Ingole, A.S. (2001).Effect of land configuration and nutrient management on yield of rainfed cotton. *J. Soils and Crops,* **11 (1)**: 125-127.

Dangore, S.T., Chaudhari, C.S., Panchabhal, P.R. and Deshpande, R.M. (2001). Effect of land configuration and nutrient management on growth and yield of rainfed cotton. *J. Soils and Crops,* **11 (2)**: 219-222.

Jogadande, V., Malvi, G.C., Dalal, S.R. and Karunkar, A.P. (2003). Effect of different layouts and nitrogen levels on growth and yield of soybean *PKV. Res. J.,* **27(2)**: 183-184.

Kachapur, M.D., Radder, G.D. and Biradar, B.N. (1981). Quality studies in niger genotypes in relation to spacing and fertility levels. *Oilseeds J.,* **11**: 38-41.

Chapter 32

Callus Induction in *Andrographis paniculata* Nees.

***V.G. Ambewadikar*[1], *R.P. Bansode*[1] *and A.B. Ade*[1]**

ABSTRACT

Andrographis paniculata Nees (Acanthaceae), commonly known as 'Kalmegh' is traditionallyused medicinal plant.The explants were excised and cultured in callus induction media which consisted of Murashige and Skoog (MS) media containing various concentrations of 2, 4- Dichlorophenoxy acetic acid (2,4-D) and Indole acetic acid (IAA) alone or in combination with benzyladeninepurine (BAP) (benzyladenine) or kinetin. After four weeks in culture, it was observed that for *A. paniculata,* the contamination rate was 57.5 per cent and the callus induction rate was 28.33 per cent. In present study callus was raised from the nodal and intermodal segments. Maximum callus was obtained on MS medium supplemented with 2, 4- D (1.5 mg/liter) and IAA (2 mg/l).

Keywords: Andrographis paniculata, Callus induction, Medicinal plant.

Introduction

Andrographis paniculata (Burm. f.) Wallich ex Nees, family Acanthaceae and known as 'Kalmegh' and distributed throughout tropical India. It is used as an herbal medicine. It is laxative, dry, cooling, bitter, and overcomes difficulty in breathing, hemopathy, burning sensation, cough, edema, thirst, skin diseases,

---

1 Plant Tissue Culture Laboratory, Department of Botany, Dr. Babasaheb Ambedkar Marathwada University, Aurangabad – 431 004, M.S., India

syphilitic cachexia, syphilitic ulcers, worms, acidity, and liver complaints (Sivarajan and Balachandran, 1994). The important compounds isolated from different parts of the plant are apigenin-7,4-di-O- methyl ether, carvacrol, eugenol, myristic acid, hentriacontane, tritriacontane, oroxylon A, wogonin, and diterpenoids like andrograpanin, andropanoside, andrographolide, and neoandrogra-pholide (Rastogi and Mehrotra, 1993). Conventional vegetative propagation of this important plant is very difficult and too slow to meet the commercial quantities required. Variability among the seed-derived progenies and scanty and delayed rooting of seedlings curb its propagation via seeds. Micropropagation through somatic embryogenesis is an option for the rapid production of uniform plants. True-to-type nature of the somatic embryo-derived plantlets has been reported (Jayanthi and Mandal, 2001; Tokuhara and Mii, 2001).

The present investigation has the impact of different concentrations of 2, 4–D and IAA on *in vitro* callus induction in *Andrographis paniculata* Nees from node, internode and leaf as explants.

Materials and Methods

In the present study, induction and culture of callus of *A. paniculata* were investigated. Young leaves were collected from field-grown plants of *A. paniculata* Nees and were washed thoroughly in running tap water, surface sterilized in 0.1 per cent (w/v) mercuricchloride for 10 min and washed three times in sterilede-ionized water. The sterilized explants were cut into 5 × 5 mm square segments and cultured onto Murashige and Skoog (MS) agar medium (0.2 per cent w/v gelrite) supplemented with 30 g/l (w/v) sucrose and indoleacetic acid (IAA; 0.5, 1.0, 1.5, 2.0 and 2.5 mg/l), 2, 4-D (0.5, 1.0, 1.5, 2.0 and 2.5 mg/l) and BAP (0.1, 0.2, 0.3, 0.4 and 0.5 mg/l). The pH of the medium was adjusted to 5.8 ± 0.2 before autoclaving (121°C for 15 min). The cultures were incubated in the dark at 25 ± 1°C.

Table 32.1: Effect of Various Plant Growth Regulators for Callus Induction in *Andrographis paniculata* from Node, Internode and Leaf Explants

Sl.No.	*Phytohormones (mg/l)*			*Explant Producing Callus*	*%of Respond*
	2,4-D	*BAP*	*IAA*		
1.	0	0	0	0	0
2.	0.5	0.1	0	2/12	16.67
3.	1.0	0.2	0	3/12	25.00
4.	1.5	0.3	0	6/12	50.00
5.	2.0	0.4	0	4/12	30.00
6.	2.5	0.5	0	5/12	41.66
7.	0	0.1	1.0	1/12	0.80
8.	0	0.2	1.5	4/12	30.00
9.	0	0.3	2.0	5/12	41.66
10.	0	0.4	2.5	3/12	25.00
11.	0	0.5	3.0	3/12	25.00

Results and Discussion

Internode and leaf explants of *A. paniculata* cultured on MS basal medium did not form callus, but explants remained viable for up to 25 days. Explants formed callus on MS medium fortified with 2,4-D (0.5-2.5 mg/l) in combination with BA (0.1–0.5 mg/l). IAA in combination with BA induced callus to varying degrees depending on the concentration and type of explants (Table 35.1). Calli grown on medium with IAA and BAP were semi-hard and pale green in color. The different levels of 2,4-D (0.5-2.5 mg/l) used, the medium with 1.50mg 2,4-D was best for the formation of callus, which was friable and creamy yellow. The combination of 1.50 mg 2,4-D and 2.00 mg BAP was superior, as friable callus was initiated within 10 days.

Acknowledgement

The authors gratefully acknowledge Dr. V. S. Kothekar, Professor and Head, Department of Botany, Dr. Babasaheb Ambedkar Marathwada University, Aurangabad for providing facilities.

References

Jayanthi, M. and Mandal, P.K. (2001).Plant regeneration through somatic embryogenesis and RAPD analysis of regenerated plants in *Tylophora indica* (Burm. f. Merril.) *In vitro* Cell. *Dev. Biol. Plant*, **37**: 576-580.

Murashige, T. and Skoog, F. (1962). A revised medium for rapid growth and bioassays for tobacco tissue cultures. *Physiol. Plant.*, **15**: 473-497

Rastogi, R.P. and Mehrotra, B.N. (1980). *Compendium of Indian Medicinal Plants*, Vol. 3 -1984. New Delhi: CDRI and Publication and Information Directorate, pp. 41-42.

Sivarajan, V.V. and Balachandran, I. (1994). *Ayurvedic Drugs and their Plant Sources.* Oxford and IBH Publishing, New Delhi, pp. 243-245.

Tokuhara, K. and Mii, M. (2001). Induction of embryogenic callus and cell suspension culture from shoot tips excised from flower stalk buds of Phalaenopsis (Orchidaceae). *In vitro* Cell. *Dev. Biol. Plant*, **37**: 457-461.

Chapter 33

Biotechnology for Mass Production of Biocontrol Agents *in vitro* Against Plant Pathogenic Fungi

S.S. Patale[1] and D.S. Mukadam[2]*

ABSTRACT

Biological control by antagonistic organism is a potential, ecofriendly, non-chemical and sustainable approach for managing plant diseases. *Trichoderma* species is being used as biocontrol agent against number of seed-borne and soil-borne diseases of crops caused by fungi. On this account commercial mass production of *Trichoderma* is a demand of the time. Therefore in the present experiments various substrates were used for growth as spoiled food grains, agricultural waste and animal dung were used for the cultivation of *Trichoderma viride*. Among six types of animal dung used maximum growth was supported by goat dung, horse dung, cow dung and buffalo dung. Similarly agricultural waste products maximum spore production of *Trichoderma viride* took place on rice husk, maize cob and jowar straw. among seeds of rice, maize, bajra, jowar and wheat (cereals) and tur, black gram, pea (legumes) and groundnut, safflower and sunflower(oilseeds) supported maximum production of *Trichoderma viride*.

1 Department of Botany, S.A.J.V.P.M[S.] Gandhi College, Kada Tq. Ashti Dist. Beed, M.S., India

2 Dr. Babasaheb Ambedkar Marathwada University, Aurangabad, M.S., India

* Corresponding Author E-mail: pataleshivraj@yahoo.co.in

Effectiveness of *T.viride, T. harzianum, T. hamatum* and *Trichoderma* ($local_1$) were screened against several plant pathogenic fungi by dual culture technique. *Trichoderma* ($local_1$) and *T. harzianum* were found to the strong antagonistic against plant pathogenic fungi than other *Trichoderma* species.

Keywords: *Agro waste, Biocontrol, Mass production, Plant pathogenic fungi, Trichoderma sp.*

Introduction

In the recent years, pollution caused by excessive use of chemical pesticides increased the interest in integrated pest management. Among the greatest hazards in crop production, diseases and insect pests are the main problems. Any one of them can upset crop yields with suddenness. For combating diseases, successful measures of chemical control have been developed over the years. Though chemicals have played a significant role in maximizing crop productivity, extensive use of broad spectrum compounds, some of which are non-degradable, has resulted in variety of harmful and undesirable effects not only on man and wild-life, but on the ecosystem as a whole. With the increasing awareness of the problems and expense of conventional methods of disease control, including fungicides and costly and time consuming breeding programmes, biological control of plant pathogens has many attractions. The term "Biological control" was first used by Smith (1919) to signify the use of natural enemies to control insect pests.

Many techniques for large scale biomass generation of this fungus are still in infancy and cost is not competitive enough for easy adaptation by commercial bioagent producers. Consequently, any media used for mass production of *Trichoderma* species must be economic and be able to support production of large quantities of biomass and viable propugles. A variety of media have been used by various researchers for production of *Trichoderma* spp. In stationary flasks (Patel *et al.*, 1994) and liquid fermenters (Papavizas *et al.*, 1984) and different substrate (Patale, 2005). Molasses brewers yeast medium is being used widely for commercial production of *Trichoderma* by fermentation process (Lumsden *et al.*, 1995). Saju *et al.* (2002) used farm waste like neem cake, coirpith, farmyard manure and decomposed coffee pulp for the production of *T. harzianum*. *Trichoderma* spp. are free living fungi that are common in soil and root ecosystems. Recent discoveries shows that they are opportunistic, avirulent plant symbionts, as well as being parasites of other fungi (Harman *et al.*, 2004).

Biological control mainly consist of using a microorganisms to control harmful microorganisms causing plant disease without disturbing the ecological balance. Weindling (1932) suggested the potential use of *Trichoderma* Species as a biocontrol agent against the soil borne plant pathogens like *Rhizoctonia solani*. The biological control of root diseases of crop plants by introduction of antagonistic microorganism has been suggested as an environmentally safer alternative to the use of fungitoxic chemicals (Baker and Cook, 1974). Somasekhara *et al.* (1996) studied under green house conditions the biological control of *Fusarium udum*, the pigeon pea wilt pathogen, by six isolates of *Trichoderma* Species when the application of *Trichoderma* species significantly reduced the number of *F.udum* propogules and wilt incidence.

Material and Methods

Mass Multiplication of *Trichoderma* spp.

In Conical Flasks

Two different categories of substrates, *viz.* agrowastes (sugarcane baggase, redgram pods, soyabean straw, greengram straw, blackgram straw, matki straw, bajra straw, wheat straw, jowar straw, rice husk and maize cobs) and crop seeds (maize, bajra, jowar, wheat, rice, redgram, blackgram, matki, pea, groundnut, mustard, hemp, sesamum, sunflower, safflower and kala til) were evaluated for growth responses and conidial yield of *T. viride* through solid state fermentation technique. For determination of conidiophore production of the bioagent on different substrates, 20 gm of each substrates 20 ml water was added in 250 ml conical flasks with cotton plugged and autoclaved at 15 lb psi for 1 hour. Flasks were inoculated with actively growing culture of *T. viride*, keeping three replications and were incubated at 25 ± 2°C for 12 days. For comparing the ability of growth and sporulation on these substrates colony forming units (CFU g^{-1}) were determined by dilution plate technique on selective medium.

In Poly Propylene Bags

Two different categories of substrates, *viz.* agrowastes (sugarcane baggase, redgram pods, soyabean straw, greengram straw, blackgram straw, matki straw, bajra straw, wheat straw, jowar straw, rice husk and maize cobs) and crop seeds (maize, bajra, jowar, wheat, rice, redgram, blackgram, matki, pea, groundnut, mustard, hemp, sesamum, sunflower, safflower and kala til) were evaluated for growth responses and conidial yield of *T. viride* through solid state fermentation technique. For determination of conidispore production of the bioagent on different substrates, 200 gm of each substrate adjusted with 70 per cent moisture content, was filled into (21 x 6) polypropylene bags fitted with staples. The bags were then sterilized by autoclaving at 15 lb psi pressure for 1 hour. One millilitre spore suspension of *T. viride* in sterilized distilled water was used to inoculate each bag with the help of disposable syringe. Inoculated bags were incubated at 25 ± 2°C for 12 days. For comparing the ability of growth and sporulation on these substrates colony forming units (CFU g^{-1}) were determined by dilution plate technique on selective medium.

Antagonistic Activity–*Trichoderma* spp. as Biocontrol Agent

Antagonistic potential of *Trichoderma* spp. against plant pathogenic fungi was studied by dual culture method (Sudhamoy Mandal *et al.*, 1999). In this method an agar disc (15 mm) containing mycelial growth of plant pathogenic fungi was inoculated at the four corners of PDA poured petriplates and culture discs of *Trichoderma* spp. were placed at center leaving 1 cm distance. Petriplates were incubated for a week at 25 ± 1°C. Plates without antagonist served as control. Two replicates were kept for each treatment and observations on colony diameter (mm) overgrowth, lysis and formation of inhibition zone were recorded to select highly effective nature of *Trichoderma* spp. Tests were also carried out to evaluate the efficacy of *Trichoderma* spp. against the post harvest pathogens (Vincent, 1927).

Results and Discussion

Mass Multiplication of *Trichoderma viride* on Different Substrates

Mass Multiplication of *Trichoderma viride* on Different Animal Dung

It is regular practice of farmers to use animal dung along with agricultural waste for the preparation of FYM (farm yard manure) therefore in order to study the effect of animal dung on *Trichoderma* growth cow dung, buffalo dung, goat dung, horse dung as well as poultry excreta were tested for the cultivation of *Trichoderma* species. The dung in the form of solid (mud) were used for the cultivation of *Trichoderma*. The results are given in Table 33.1.

Table 33.1: Mass Multiplication of *Trichoderma viride* on Animal Dung (12 day ncubation)

Animal Dung	*c.f.u.g^{-1} (1 x 10^{-5})*
Cow dung	265
Buffalow dung	195
Goat dung	277
Poultry	120
Horse dung	280

It is clear from the results given in Table 33.1 that horse dung, goat dung and cow dung supported maximum biomass of *Trichoderma* while poultry excreta proved to be poor source.

Mass Multiplication of *Trichoderma viride* on Different Crop Seeds (Substrates)

In order to find out an economical viable substrate for the large scale production of *Trichoderma* as a bio-control agent, seeds of sixteen different crops were used for the cultivation of *Trichoderma*. Seeds sterilized in plastic bags inoculated with spore suspension of *Trichoderma viride* were incubated at 25 ± 2°C temperature for 12 days. It is clear from the results given in Table 33.2 that seeds of maize, bajra, and rice among cereals, redgram, blackgram and pea among legumes, and groundnut among oilseeds supported maximum growth as well as sporulation of the fungus, while seeds of mustard, sesamum and hemp proved to be poor substrates for mass multiplication of *Trichoderma*.

Table 33.2: Mass Multiplication of *Trichoderma viride* on Different Farm Wastes

Farm Waste Substrates	*c.f.u.g^{-1} (1 x 10^{-5})*
Sugarcane baggase	40
Redgram Pod	180
Soyabean straw	60
Blackgram straw	80
Greengram straw	75
Matki straw	30
Bajra straw	35
Ricehusk	210
Wheat straw	30
Maize cob	200
Jowar straw	130

Mass Multiplication of *Trichoderma viride* on Different Farm Wastes

A large number of agricultural wastes are available and continuously produced due to the crop harvest. Therefore 20 gm of agriculture waste moist with 20 ml of water was taken individually in the conical flask. The flasks were autoclaved and the substrates were used for growth of *Trichoderma viride*. After 12 days of incubation at room temperature the spore concentration was measured. It is clear from the results given in Table 33.3 that among 11 different types of agriculture wastes used for mass multiplication and rice husk, maize cobs and red gram pods supported maximum sporulation and growth of the fungus and sugarcane baggase, matki straw, bajra straw proved poor substrate.

Table 33.3: Mass Multiplication of *Trichoderma viride* on Different Substrates (Crop seeds)

Crop Seeds	*CFUg⁻¹ (1 x 10⁻⁵)*
Maize (*Zea mays*)	212
Bajra (*Pennisetum typhoides*)	215
Jowar (*Sorghum vulgarae*)	205
Wheat (*Triticum aestivum*)	212
Rice (*Oryza sativa*)	260
Red gram (*Cajanus cajan*)	210
Blackgram (*Phaseolus angularis*)	200
Matki (*Phaseolus aconitifolius*)	150
Pea (*Pisum sativum*)	140
Groundnut (*Arachis hypogaea*)	180
Mustard (*Brassica campenstris*)	90
Hemp (*Hibiscus cannabinus*)	80
Sesamum (Til) (*Sesamum indicum*)	130
Sunflower (*Helianthus annuus*)	120
Safflower (*Carthamus tinctorius*)	90
Kala til (Karala) (*Guizotia abyssinica*)	125

Antagonistic Activity of *Trichoderma* spp.

Four different species of *Trichoderma* were tested against 13 plant pathogenic fungi. The antagonistic effect was tested by dual culture method and results are given in the Table 33.4.

It is clear from the result summarized in Table 33.4 that all the plant pathogens were found to be significant inhibition for the growth in the presence of *Trichoderma* spp. It was also interesting that *T.harzianum* and *Trichoderma* species (local$_1$) isolated from this region proved to be stronger antagonistic as compared to other species of *Trichoderma*.

It was observed that the possible mechanism of bioagents in controlling plant pathogenic fungi. This supports earlier investigations by Roy and Sayre (1984), Chet and Inbar(1994), Somasekhara *et al.* (1996), Prasad and Rangeshwaram (2000) and Patale(2005). In the present study it also clearly evident that antagonistic effects of all the *Trichoderma* Spp. against plant pathogenic fungi.

Table 33.4: Antagonistic Effect of Different *Trichoderma* spp. Against Plant Pathogenic Fungi (results after 7 days)

Plant Pathogenic Fungi	*Control (Without Trichoderma sp.) (mm)*	*Per cent Inhibition of Growth Due to*			
		T. viride	*T. harzianum*	*T. hamatum*	*Trichoderma (Local,)*
Aspergillus niger	90.00	74.44	86.67	80.00	88.89
A. flavus	75.00	66.67	72.00	73.00	77.33
A. parasiticus	60.00	56.67	70.00	65.00	66.67
F. oxysporum	80.00	61.25	66.25	70.00	68.75
F. roseum	85.00	62.35	67.06	70.59	69.41
Rhizoctonia solani	70.00	62.86	65.71	71.43	68.57
Penicillium notatum	75.00	57.33	80.00	62.67	76.00
Phytophthora sp.	50.00	38.00	56.00	46.00	56.00
Helminthosporium sp.	80.00	58.75	61.25	62.50	62.00
Alternaria solani	75.00	65.33	60.00	58.67	56.00
Curvularia lunata	80.00	57.50	56.25	68.75	57.00
Rhizopus stolonifer	70.00	52.86	67.14	71.43	65.00

References

Baker, K.K. and Cook, R.J. (1974). *Biological Control of Plant Pathogens*. Freeman, San Franscisco, pp. 433.

Chet, I. and Inbar, J. (1994). Biological control of fungal pathogens. *Appl. Biochem. Biothchnol.*, **48**: 37-43.

Harman, G.E., Charles R. Howell, Adá Viterbo, Ilan Chet and Matteo Lorito (2004). *Trichoderma* species opportunistic, Avrivulant plant symbionts. *Microbiology*, **2**: 43-60.

Lumsden, R.D., Lewis, J.A. and Fravel, R.D. (1995). Bioradiational pest control agents. In: *Formulations and Delivery*, (Eds.) Franklin, R. Hall and John W. Barry. ACS Symposium series 595. American Chemical Society, Washington, pp. 306.

Papavizas, G.C., Dunn, M.T., Lewis, J.A. and Beagle-Restaino, J. (1984). *Phytopathology*, **74**: 1171-1175.

Patel, S.T., Mishra, Ashok and Mishra, A. (1994). *Gujrat Agril. Univ. Res. J.*, **19**: 53-56.

Patale, S.S. (2005). Studies on *Trichoderma* for the biocontrol of plant pathogens. *Ph.D. Thesis*, Dr. Babasaheb Ambedkar Marathwada University, Aurangabad(M.S.) India.

Prasad, R.D. and Rangeshwaram, R. (2000). Effect of soil application of a granular formulation of *Trichoderma* and damping-off of Chickpea. *J. Mycol. Pl. Pathol.*, **30(2)**: 216–220.

Roy, A.K. and Sayre, R.M. (1984). Controlling sheath blight in Rice. *Louisiana Agriculture*, **27**: 16-18.

Saju, K.A., Anandraj, M. and Sarma, Y.R. (2002). On farm production of *Trichoderma harzianum* using organic matter. *Indian Phytopath.*, **55(3)**: 277-281.

Smith, N.S. (1919). On same aspects of insect control by the biological method. *J. Econ. Entomol.*, **12**: 288.

Somasekhara, Y.M., Anilkumar, T.B. and Siddaramaiah, A.L. (1996). Biocontrol of pigeonpea [*Cajanus cajan* (L.) Millsp] wilt (*Fusarium udum* Butler). *Mysore J. Agric. Sci.*, **30(2)**: 163-165.

Sudhamoy, Mandal, Srivastava, K.D., Aggarwal, Rashmi and Singh, D.V. (1999). Mycopaeasitic action of some fungi on spot blotch parhogen (*Drechslera Sorokiniana*) on wheat. *Indian Phytopath.*, **52(1)**: 39-43.

Weindling, R. (1932) *Trichoderma lignorum* as a parasite of other fungi. *Phytopathology*, **22**: 837.

Vincent, J.M. (1927). Distortion of fungal hyphae in the presence of certain inhibitors. *Nature*, **159**: 850.

Chapter 34

Air Pollution Tolerance Index (APTI) of Some Avenue Plants

R.B. Allapure[1]* and R.M. Kadam[2]

ABSRACT

Air pollution tolerance index (APTI) of 30 species growing adjacent to roadside and exposed to varying degrees of air pollution were determined by calculating ascorbic acid, total chlorophyll, leaf extract pH and relative water content of tissues. It is found that following top five plant species in urban areas which contribute rapid amelioration of polluted habitat and serve as sink to air pollution are *Ficus benghalensis* L., *Azadirachta indica* A. Juss., *Terminalia catappa* L., *Madhuca indica* Gmel., *Eucalyptus maculata* Hook. and *Tectona grandis* L.f. these plants can be used as bioindicators and also be grown as bioaccumulators.

Keywords: APTI, Biochemical changes, Avenue plants.

Introduction

Emissions of gaseous air pollutants have increased in the last decade in spite of increased controls and concern for air quality. Prediction of future development also indicates that a further increase in emission must be expected. When it is considered that total control of air pollutants is technically and especially economically

---

1 Department of Botany, M.U. Mahavidylaya, Udgir Dist. Latur, M.S., India

2 Department of Botany, M.G. Mahavidyalaya, Ahemedpur Dist. Latur, M.S., India

* Corresponding Author E-mail: allapure@yahoo.co.in

impossible, it is important that, in the future, emissions are controlled within a technical and economic framework to such an extent that ambient pollutant concentrations near the ground present no hazard to man or his environment.

Compared with man, animals, or materials, plants respond very sensitively to widely distributed pollutants, such as sulphur dioxide, hydrogen fluoride (HF) and hydrogen chloride. Extensive loss to agriculture and lasting changes in natural ecosystem are the result.

Studies on the effect of air pollution on vegetation, therefore, provide an important basis, particularly for preventive measures in air pollution control. Responses of plants to pollutants are not only primarily dependent on pollutant concentration and exposure time, but also on the amount of pollutant absorbed by the per unit of time.

Keeping in view of the increasing importance of bio-indicators in environmental monitoring. Generally the plant responses to pollutants are characteristic rather than specific. Attempts have been made to assess certain plant species which can specifically categorized as sensitive for a particular pollutant. The sensitive species were used for monitoring of air pollutant. The exposed plants are then analyzed with respect to changes in the selected parameters. The changes are then compared to those obtained in the same species exposed at pollution free zone and taken as control.

Materials and Methods

Thirty different angiosperms plants were selected to determine Air Pollution Tolerance Index (APTI).

Air pollution tolerance index (APTI) shows the tolerance level of a plant to air pollution. Singh and Rao (1983) made an attempt to determine the APTI values, to get an empirical value for the tolerance level of a plant to air pollution. The formula suggested by them was as follows:

$$APTI = \frac{A(T+P)+R}{10}$$

where, A is ascorbic acid content of leaf in mg/gm dry weight, P is the leaf extract pH, T is total chlorophyll content and R is the relative water content of leaf. The entire sum was divided by 10 to obtain a small manageable figure.

Result and Discussion

APTI (Air Pollution Tolerance Index)

The APTI values have been investigated for many areas. This value gave a correct picture of the tolerance/sensitivity level of plants (Table 34.1). The value were highest (56.20) for *Tectona grandis* L.f. and lowest (21.55) for *Parkia biglanduloa* Wt. and Arn. in the industrial area of Udgir. APTI values in response to automobile pollution at Udgir showed minimum reduction in *Azadirachta indica* A. Juss. and maximum reduction in *Parkia biglanduloa* Wt. and Arn. Therefore, the former was seen as highly tolerant to automobile pollution while the later was observed to be the most sensitive one.

Table 34.1: Air Pollution Tolerance Index (APTI) of Some Selected Plant Taxa Located Study at Site

Sl.No.	Name of the Plant	Ascorbic Acid Content (mg/g)	Total Chlorophyll Content (mg/g)	pH	Relative Water Content %	APTI
1.	*Ailanthus excelsa* Roxb.	45	1.7	6.9	65	45.20
2.	*Albizzia lebbeck* (L.) Willd.	27.1	0.9	9.2	61.9	33.56
3.	*Annona reticulata* L.	47.5	1.9	7	70	49.28
4.	*Annona squamosa* L.	47	2	7	67	49.00
5.	*Azadirachta indica* A. Juss.	48.12	1.98	6.8	85.14	50.76
6.	*Cassia siamea* Lamk.	43	1	6.8	69.75	40.52
7.	*Dalbergia sisso* Roxb.ex Dc.	29.68	0.95	5.4	61.18	24.96
8.	*Delonix elata* (L.) Gamble.	36.41	1.1	6.8	73	36.06
9.	*Delonix regia* (*Boj.ex Hook.*) Raf.	29	2.31	6.3	81.5	33.12
10.	*Derris indica* (Lamk.) Bennett.	45.5	0.74	6.8	53.2	39.63
11.	*Erythrina indica* Lamk.	51	1.5	7	63.2	49.67
12.	*Eucalyptus maculata* Hook.	62	1.4	6.5	60	54.98
13.	*Ficus benghalensis* L.	47.45	2.31	6.5	86.2	50.42
14.	*Ficus hispida* L.f.	41	1	7.5	39	38.75
15.	*Ficus religiosa* L.	46.3	2	6.9	85.17	49.72
16.	*Madhuca indica* Gmel.	57	1.5	6.9	60	53.88
17.	*Mangifera indica* L.	27.68	1.68	5.9	75.9	28.57
18.	*Millingtonia hortensis* L.f.	45	1.6	6.6	60	42.90
19.	*Mimusops elengi* L.	43.1	0.9	6.5	74	39.29
20.	*Morinda citrifolia* L.	42	1.15	7	70	41.23
21.	*Moringa oleifera* Lamk.	28	1.1	5.5	75	25.98
22.	*Parkia biglanduloa* Wt. and Arn	17.2	2.25	6.5	65	21.55
23.	*Peltophorum pterocarpum* (DC.) Baker	18	2.25	6.5	68	22.55
24.	*Polyalnthia longifolia* S (Sonner)Thw.	23.71	0.9	9.8	39.45	29.31
25.	*Spathodia companulata* P.Beauv.	57	1.25	6.5	52	49.38
26.	*Sterculia foetida* L.	37.18	2.18	5.9	77.9	37.83
27.	*Tamarindus indica* L.	22.18	0.65	6.8	62.8	22.80
28.	*Tectona grandis* L.f.	65	1.4	6	81	56.20
29.	*Terminalia catappa* L.	50.1	2.1	6.7	68	50.89
30.	*Thespesia populnea* (L.) Soland.ex Corr.	22.5	1	6.5	49	21.78

Observations (Table 34.1) on the estimation of APTI values of plants growing in the Heavy Traffic site (HTS) I of Udgir town indicate that it was least in *Parkia biglanduloa* Wt. and Arn while it was highest in Heavy Traffic site (HTS) II for *Tectona grandis* L.f. further, air pollution tolerance level of each plant was different and different species showed different behaviour. The plant responses to pollutants thus depend upon a number of factors. It is seen that plants having higher index values are more tolerant to air pollution and can be used as a filter or sink to mitigate pollution, while plants having low index value can be used to indicate levels of air pollution. (Agarwal and Bhatnagar, 1982)

The calculated APTI in control plants were higher than those recorded for polluted sites, and decreased progressively with increase in traffic density from 56.20 to 21.55 (Table 34.1). Air pollution tolerance index in *P. biglanduloa* Wt. and Arn showed maximum reduction at HTS I. The mean per cent reduction of APTI over control (per cent ROC) demonstrated a sharp decline in *T. populnea* (L.) Soland.ex Corr., *M. oleifera* Lamk., *P. biglanduloa* Wt. and Arn, *P. pterocarpum* (DC.) Baker and *P. longifolia* S (Sonner) Thw. While less significant changes were apparent in *A. indica* A. Juss., *F. benghalensis* L., *F. religiosa* L., *M. citrifolia* L. and *T. catappa* L..

The observed APTI value 49.72 in *Ficus religiosa* L. agrees closely with the report of Datta and Ray (1995). However, due to marginal reduction in APTI (2.62 per cent) in *A. indica* A. Juss.the conclusion of Sunita and Rao (1997) that it is sensitive to air pollution does not appear to be true in our study. In addition neither chlorophyll nor ascorbic acid levels decreased significantly in this case.

Datta and Ray (1995) concluded that species having low index values are more sensitive to air pollution and *vice versa*. In the context of the present findings, this appears to be an arbitrary classification because the level of total chlorophyll ascorbic acid, relative water content and pH; which determine APTI level of a species; are intrinsic features of each species and no comparisons can be made amongst the species.

Due to higher reduction in air pollution tolerance index over their control counterparts *M. oleifera* Lamk., *P. biglanduloa* Wt. and Arn, *P. pterocarpum* (DC.) Baker, *P. longifolia* S (Sonner) Thw. and *T. populnea* (L.) Soland.ex Corr. are considered as relatively sensitive species. Conversely *A. indica* A. Juss., *F. benghalensis* L., *F. religiosa* L., *M. citrifolia* L. and *T. catappa* L. are considered as relatively resistant species due to their least or insignificant reductions in APTI. The less degradation of chlorophyll and ascorbic acid contents further substantiates their resistant abilities. It is noteworthy that the plants of the former category are regarded as ideal species, which could be effectively employed for phytomonitoring automobile exhaust pollution along side the busy traffic ways.

Air pollution tolerance level of each plant is different, and plants do not show a uniform behaviour. Plants on the basis of their response to pollutants under field conditions and laboratory conditions have been classified into sensitive and tolerant species (Jacobson and Hill, 1970). The degree of sensitivity of a plant depends on its development stage, nutritional status and other ecological factors (Guderian, 1977). Many other factors such as stomatal resistance to the entry of gases have been held responsible for the expression of pollution response. It is a universal logic that stress can either be avoided or tolerated through physiological manipulation to toxic pollutants entering into the plant body (Levitt, 1980).

The APTI determination involves assessment of the plant parameters as per their relative significance. Singh and Rao (1983) have assigned more importance to foliar ascorbic acid, as it is multiplied to the sum of total chlorophyll and leaf extract pH. Ascorbic acid is a strong reductant. It activates many physiological and defense mechanisms and its reducing power has been known to be directly proportional to its concentration. It also influences resistance to adverse environmental conditions, including air pollution (Keller and Schwager, 1977).

A plant species known to be sensitive or tolerant in one geographical area might not be so in another geographical area. The external ecological factors play their role in determination of sensitivity levels of plants because the internal physiological conditions of a plant depend much upon the external ecological factors. Moreover, tolerance or sensitivity to air pollution is a reflection of internal physiological conditions of the plant. The sensitivity level of plants to air pollution is different for herbs, shrubs and trees. Trees and shrubs having identical index for tolerance level may not show similar behaviour. Therefore, plants on the basis of their habit should be assessed separately for their pollution tolerance level.

Therefore, the APTI bio analysis is useful for the identification of suitable biomonitoring (phytomonitors) for polluted urban environment. The species growing in such hostile roadside environment present the best material to ascertain the levels of sensitivity, tolerance and resistance. Raising such tolerant species in polluted habitats will lead to rapid amelioration of habitat to cope up with polluted environment. Such plants are shown to be effectively used as indicators of pollutant scavengers and serve as sink to air pollutants. The top five species in urban area which contribute maximum pollutants stocking are *Ficus benghalensis* L., *Azadirachta indica* A. Juss., *Terminalia catappa* L., *Madhuca indica* Gmel., *Eucalyptus maculata* Hook. and *Tectona grandis* L.f. (Table 34.1).

References

Agarwal, S. K. and Bhatnagar, D.C. (1982). Auto-vehicular air pollution induced pigment and ascorbic acid changes in avenue plants. *Acta Ecol.*, **13(1)**: 1-4.

Datta, S.C. and Sinha Ray, S. (1995). APTI value of five different species of higher plants from the metropolis of Calcutta. In: *Environmental and Adoptive Biology of Plants*, Scientific Publishers, Jodhpur, pp. 289-294.

Guderian, R. (1977). Air pollution: Phytotoxicity of acidic gases and its significance. In: *Air Pollution Control* Vol. 22, Edi. Springer-Verlag, New York, Berlin, pp. 11-22.

Jacobson, J.S. and Hill, A.C. (1970). Recognition of air pollution injury to vegetation: A pictorial atlas. Air Pollution Control Association, Pittsburg, Pennsylvania, pp. 80-81.

Keller, T. and Schwager, H. (1977). Air pollution and ascorbic acid. *Eur. J. Forest Pathol.*, **7(6)**: 338-350.

Levitt, J. (1980). *Responses of Plants to Environmental Stresses*. Academic Press, London, New York, pp. 182-184.

Singh, S.P. and Rao, D.N. (1983). Evaluation of plants for their tolerance to air pollution. *Proc. Symp. Air Pollution Control*, pp. 218-224.

Sunita, M. and Rao, KV.M. (1997). Air pollution tolerance capacities of selected plant species. *J. Indian Bot. Soc.*, **76**: 95-98.

Chapter 35

Effect of *Azospirillum* on Biometric Characters of Irrigated Cotton

M.B. Patil[1]*, *S.M. Kadam*[2] *and B.W. Somwanshi*[3]

ABSTRACT

A field experiment was conducted during the year 2003-2004 to study the effect of *Azospirillum* on Biometric characters and yield of irrigated cotton. The field experiments were conducted using Randomized Block Design (RBD). Amongst the different strains of *Azospirillum*, Surat Strain of *Azospirillum* brought about maximum increment in shoot and root length, shoot and root weight (dry and fresh) on all the dates of observations. It was followed by HAU Strain of *Azospirillum*.

Keywords: Azospirillum, Biometric characters, Cotton.

Introduction

In India, cotton is grown in almost all the states. The productivity of cotton in our country is quite low as compared to world average (Annonymous, 2002). Therefore

---

1 L.R.W. ACS College, Sonpeth, Dist. Parbhani, M.S., India

2 Research Assistant, Soil Scientist Unit, Basmatnagar, Dist. Parbhani, M.S., India

3 Govt. Jr. College, Jorawandi, Dist. Gadchiroli, M.S., India

* Corresponding Author E-mail: mukundrajbpatil@gmail.com

efforts are to be made to increase the productivity by supplying major nutrients with the use of biofertilizers. In non-legume crops like cotton, associative symbiotic N-fixer like *Azospirillum* is known to fix substantial quantities of nitrogen, is of much importance. *Azospirillum* inoculation is equivalent to applying 40kg nitrogen per hectare. Seeds inoculated with *Azospirillum* resulted 1.5 times increase in the length of *Flax* fibre (Mikhailouskasa, 2006). Seed yield of *Brassica juncea* (L) was increased considerabely due to application of *Azospirillum* along with neem cake (Khan *et al.*, 2010). Taking into account the beneficial role of N-fixer, the present study was undertaken to access the impact of their inoculation on the biometric characters and yield of irrigated cotton.

Materials and Methods

The field experiment was carried out at Research farm, Department of Agronomy, College of Agriculture, Marathwada Agriculture University, Parbhani. Randomized Block Design was used for the experiment. Three most efficient strains (TNAU, HAU and Surat) of *Azospirillum* were used for inoculation of seed. The variety of cotton PH 348 was used for sowing. Sowing was done by dibbling method at row to row 60cm and plant to plant 30cm spacing. Irrigation was given as per the requirement of crop. Standard plant protection schedule was followed to protect the crop from diseases and pests. The observations on different growth parameters like shoot length and weight, Root length and weight were recorded on 45, 90 and 135 days after sowing (DAS). The fresh weight of root and shoot was measured physically on top loading balance and resulting weight were recorded as root and shoot fresh weight in gm. The dry matter accumulation by root and shoot recorded by subjecting the root and shoot to the oven drying at 60°C for constant weight.

Result and Discussion

The effect of inoculation of *Azospirillum* on biometric characters and yield on irrigated cotton were studied.

Length and Weight of Shoot

The effect of inoculation of *Azospirillum* on shoot length and weight of cotton was recorded at 45, 90 and 135 days after sowing in a field trial. The relevant data so obtained was presented in Table 35.1. The data clearly indicated that shoot length and weight of cotton was significantly increased with the inoculation of *Azospirillum* as compared to non-inoculated control on all dates of observation. Among the different strains of *Azospirillum*, Surat strain brought about maximum increment in shoot length (54.53, 118.83 and 134.26cm on 45, 90 and 135 DAS respectively) and dry shoot weight (25.64, 86.72 and 226.76gm on 45, 90 and 135 DAS respectively). It was followed by the HAU strain of *Azospirillum*. The results obtained in the present investigation are in agreement with those reported in the past. Mehandle (1981) obtained more plant biomass of *Sorghum* with the inoculation of *Azospirillum*. Martin *et al.* (2008) recorded 12.9 per cent increase in shoot matter inoculation in wheat plant due to inoculation of *Azospirillum*.

Table 35.1: Effect of *Azospirillum* on Shoot Length (cm) and Shoot Weight (gm) of Irrigated Cotton

Treatments	Shoot Length (cm) at DAS			Shoot Weight (gm) at DAS					
				Fresh Weight			Dry Weight		
	45	90	135	45	90	135	45	90	135
Inoculation of TNAU	52.93	113.20	128.56	45.72	228.60	525.30	22.36	76.15	211.10
Inoculation of HAU	54.15	116.44	130.79	46.80	247.00	545.70	24.18	82.02	217.60
Inoculation of Surat	54.53	118.83	134.26	49.01	264.25	574.26	25.64	86.72	226.76
No Inoculation	50.03	109.09	122.55	41.05	209.68	511.25	17.14	69.44	203.20
SE	0.31	0.94	0.46	0.31	2.17	2.35	0.37	0.60	0.61
CD at 5%	0.96	2.90	1.43	0.95	6.69	7.25	1.14	1.87	1.88

Table 35.2: Effect of *Azospirillum* on Root Length (cm) and Root Weight (gm) of Irrigated Cotton

Treatments	Root Length (cm) at DAS			Root Weight (gm) at DAS					
				Fresh Weight			Dry Weight		
	45	90	135	45	90	135	45	90	135
Inoculation of TNAU	29.45	50.34	60.05	4.35	25.21	39.50	2.31	10.90	19.01
Inoculation of HAU	30.75	51.14	61.42	4.48	27.94	41.49	2.64	12.76	19.66
Inoculation of Surat	32.02	52.41	62.84	4.61	30.46	44.08	2.93	14.60	22.33
No Inoculation	28.48	45.23	54.53	4.07	17.24	32.34	2.03	9.06	14.07
SE	0.31	0.25	0.07	0.04	0.17	0.12	0.025	0.023	0.47
CD at 5%	0.96	0.77	0.20	0.13	0.50	0.36	0.079	0.072	1.47

Length and Weight of Root

Root weight and length of cotton as influenced by inoculation of *Azospirillum* was recorded on 45, 90 and 135 DAS. The data was presented in Table 35.2. The data (Table 35.2) clearly indicated that root weight and length of cotton was significantly increased with the inoculation of *Azospirillum* as compared to non-inoculated control on all the dates of observation. The Surat strain of *Azospirillum* brought about maximum increment in root length (32.02, 32.41 and 62.84cm on 45, 90 and 135 DAS respectively). Fresh root weight (4.61, 30.46 and 44.08gm on 45, 90 and 135 DAS respectively) and dry root weight (2.93, 14.60 and 22.33gm on 45, 90 and 1 35 DAS respectively). It was followed by HAU strain of *Azospirillum*. The increase in root biomass of cotton obtained in present investigation is in full agreement with those reported in the past (Mehandale, 1981; Wange and Ranawade, 1997). Anonymous (2002) reported significant increase in root biomass of cotton when inoculated with *Azospirillum* + PSB + PPFM. Mikhailouskaya (2006) observed 1.3-1.6 times higher root mass of flax plant inoculated with *Azospirillum brasilense* than noninoculated plants. Martin *et al.* (2008) recorded 22 per cent increase in dry matter accumulation in root of wheat plant when inoculated with *Azospirillum brasilense.*Khan *et al.* (2010) observed increase in fresh as well as dry weight when *Brassica juncea* (L) seeds inoculated with *Azospirillum brasilense* than non-inoculated plants.

References

Anonymous (2002). Annual progress report of technology mission on cotton, mini-mission for the year 2002, pp. 206-219.

Khan, Irfan, Aquil, Ahmad and Anwar, Masood (2010). Effect of *Azospirillum* inoculation and organic manure on *Brassica juncea* (L.). *Int. J. of Plant Sciences,* **5(2)**: 669-671.

Martin, Diaz-Zoritaa Maria, Virginia, Fernandez, Conigiab (2008). Field performance of a liquid formulation of *Azospirillun brasilense* on dry land wheat productivity. *Eur. J. Biol. Doi.,* **10**: 1016.

Mehandale, R.K. (1981). Occurrence, Characterization and screening of nitrogen fixing *Azospirillum. M.Sc. (Agri.) Thesis,* Mahatma Phule Agri. Uni. Rahuri, M.S.

Mikhailouskaya, N. (2006). The effect of *Azospirillum brasilense* on flax yield and it's quality. *Plant Soil Environ.,* **52(9)**: 402-406.

Wange, S.S. and Ranawade, D.B. (1997). Effect of microbial inoculants on fresh root development of grape var. Kishmis chroni. *Recent Hort.,* **4**: 27-31.

Chapter 36

Evaluation of Some Antifungal Plant Extracts Against *Stemphylium vesicarium*, Inciting Leaf Blight of Onion

A.P. Suryawanshi[1], *K.T. Apet*[1], *D.N. Dhutraj*[1], *S.H. Khade*[1], *D.B. Gawade*[1] *and A.L. Harde*[1]

ABSTRACT

Aqueous extracts of 12 botanicals/plant species were evaluated *in vitro* (@10, 15, 20 per cent) against *Stemphylium vesicarium*, inciting leaf blight of onion (*Allium cepa* L.). All the botanicals tested were found fungitoxic and significantly inhibited mycelial growth of *S. vesicarium* over untreated control. However, percentage inhibition of the test pathogen was increased with increase in concentration of the botanicals extract. Among the botanicals evaluated, *Azadirachta indica* (Neem) was found most fungi toxic and recorded significantly highest mean growth inhibition (83.94 per cent) of the test pathogen. The second and third best botanicals in respect of fungitoxicity were *Oscimum sanctum* (Tulsi) and *Bougainveillia spectabilis* (*Bougainveillia*) which recorded next best maximum mean growth inhibition, respectively of 69.04 and 68.16 per cent of the test pathogen. This was followed by the botanicals *viz., Lawsonia innermis* (66.38 per cent), *Ipomea carnea* (65.90 per cent), *Zingiber officinale* (59.18 per cent), *Eucalyptus globules* (58.58 per cent),

---

1 Department of Plant Pathology, College of Agriculture, M.K.V., Parbhani – 431 402, M.S., India

Annona squamosa (56.89 per cent), *Tagetes erecta* (55.18 per cent) and *Parthenium hysterophorus* (54.96 per cent). Comparatively, *Allium sativum* (Garlic) was found least fungitoxic and recorded minimum meant growth inhibition (27.29 per cent). Thus, those botanicals proved antifungal/fungitoxic and significantly inhibited growth of *Stemphylium vesicarium* in present study, needs further confirmation of their fungitoxicity potential and then could be exploited under field conditions for eco-friendly and economical management of *Stemphylium* leaf blight of onion as well as other foliar diseases of the crop plants.

Keywords: *Botanicals, Fungitoxic, Inhibition, Leaf blight, Stemphylium vesicarium.*

Introduction

The onion (*Allium cepa* L.) a bulbus spice ice (originated from Middle East Asia and introduced in India from Palestin) vegetable crop is grown round the year throughout Maharashtra under varied agro–climatic conditions and the area under onion is increasing day by day. However, the average productivity and production of onion in the state of Maharashtra is comparatively less than that of the country's average productivity and production. Because several biotic and abiotic constraints are responsible for low yields of onion. Of the biotic constraints, diseases induced by the plant pathogens *viz.*, fungi, bacteria, viruses, nematodes etc. are the major one. The two major fungal diseases *viz.*, Purple blotch [*Alternaria porri* (Ellis)] and *Stemphylium* leaf blight [*S. vesicarium* (Wallr.) E. Simmons] are the most destructive causing heavy yield losses. During recent past years both the fungal diseases are occurring in devastating proportions under favorable environmental conditions, particularly during *Kharif* and *Rabi* seasons.

Typical symptoms (Plate 36.1A) of the disease appeared on foliage and foliage sheath are: small, water soaked lesions of light yellow to brown coloured. As the lesions expand, they coalesce and cause extensive blightening of the leaves. Typically lesions are found in higher number on the sides of leaves facing wind. The centers of lesions turn brown, due to abundant sporulation of the pathogen, sometimes fruiting bodies called 'perithecia' may appear in infected tissues as small, black, pinhead-like raised bodies.

Foliage blight caused by *S. vesicarium* (Wallr.) E. Simmons has been reported to inflict heavy yield losses to the tune of 80-90 per cent (Gupta, 1986; Gupta and Srivastava, 1988; Barnwal, *et al.*, 2003; Shahanaz *et al.*, 2007) in the onion crop. Though, number of fungicides were reported (Amresh and Nargund, 2004; Kumari *et al.*, 2006; Kharakwar *et al.*, 2006; Khosla *et al.*, 2007) effective against *Stemphylium* leaf blight of onion; but their indiscriminate use has posed the problems of environmental pollution, residual toxicity, development of resistance and health hazards to human and animals. Hence, for sustainable agriculture and conservation of the environment; eco-friendly, economical and effective plant disease management strategies needs to be developed and exploited commercially. A number of plant species/plant base byproducts have been reported possesses antifungal properties.

Therefore, considering seriousness of the disease and undesirable effect of the agro-chemicals, an attempt was made to evaluate the antimicrobial/antifungal potentials of locally and commonly available botanicals/plant species against *Stemphylium vesicarium*, causing leaf blight of onion. The study was undertaken at the Department of Plant Pathology, College of Agriculture, Lattur during *Kharif* and *Rabi*, 2009.

Materials and Methods

Aqueous extracts of 12 botanicals *viz*., Neem (*Azadirachta indica*), Marigold (*Tagetes erecta*), Tulsi (*Oscimum sanctum*), Eucalyptus (*Eucalyptus globulus*), Bougainveillia (*Bougainveillia spectabilis*), Mehandi (*Lawsonia innermis*), Custard apple (*Annona squamosa*), Ginger (*Zingiber officinale*), Periwinkle (*Vinca rosea*), Parthenium (*Parthenium hysterophorus*), Beshrum (*Ipomea carnea*), and Garlic (*Allium sativum*) were evaluated (@10 per cent, 15 per cent and 20 per cent) *in vitro* against *S. vesicarium*. Leaf extracts were prepared by grinding with mixture-cum grinder the 100 gm washed leaves, Ginger rhizomes and Garlic bulbs seperately in 100 ml distilled water (w/v) and filtered through double layered muslin cloth. The filtrates obtained were further filtered through Whatman No. I filter paper using funnel and volumetric flasks (100 ml cap.). The final clear extracts/filtrates obtained formed the standard plant extracts of 100 per cent concentration, which were individually evaluated, applying Poisoned food technique (Nene and Thapliyal, 1993) and using Potato dextrose agar (PDA) as basal culture medium.

An appropriate quantity of each plant extract (100 per cent) was separately mixed thoroughly with PDA medium in conical flasks (250 ml cap.) to obtain desired concentrations (10, 15 and 20 per cent) and autoclaved at 15lbs/inch^2 pressure for 15 to 20 minutes. Sterilized and cooled PDA amended with plant extracts was then poured (15 to 20ml/plate) into sterile glass Petri plates (90 mm) and allowed to solidify at room temperature. Upon solidification of PDA, all the plates were aseptically inoculated by placing in the centre a 5 mm mycelial disc obtained from a week old actively growing pure culture of *S. vesicarium*. (Plate 36.1B) Plate containing plain (without plant extract) PDA and inoculated with mycelial disc of the test fungus served as untreated control. All these plates were then incubated at 24± 1°C temperature for a week or till the untreated control plates were fully covered with mycelial growth of the test fungus. The experiment was designed in CRD and all the treatments were replicated thrice.

Observations on radial mycelial growth/colony diameter of the test fungus were recorded treatment–wise at 24 hours interval and continued till growth of the test fungus was fully covered in the untreated control plate. The observations recorded were averaged and percentage inhibition of mycelial growth over untreated control was calculated by applying the formula given by Vincent (1927).

Results and Discussion

Radial Mycelial Growth/Colony Diameter

Results (Table 36.1 and Figure 36.1) revealed that all the botanicals/plant extracts tested exhibited a wide range of growth of the test pathogen (Plate 36.2), depending

Plate 36.1A: Typical Symptoms of *Stemphylium* Blight on Onion

Plate 36.1B: Pure Culture of *Stemphylium vesicarium*

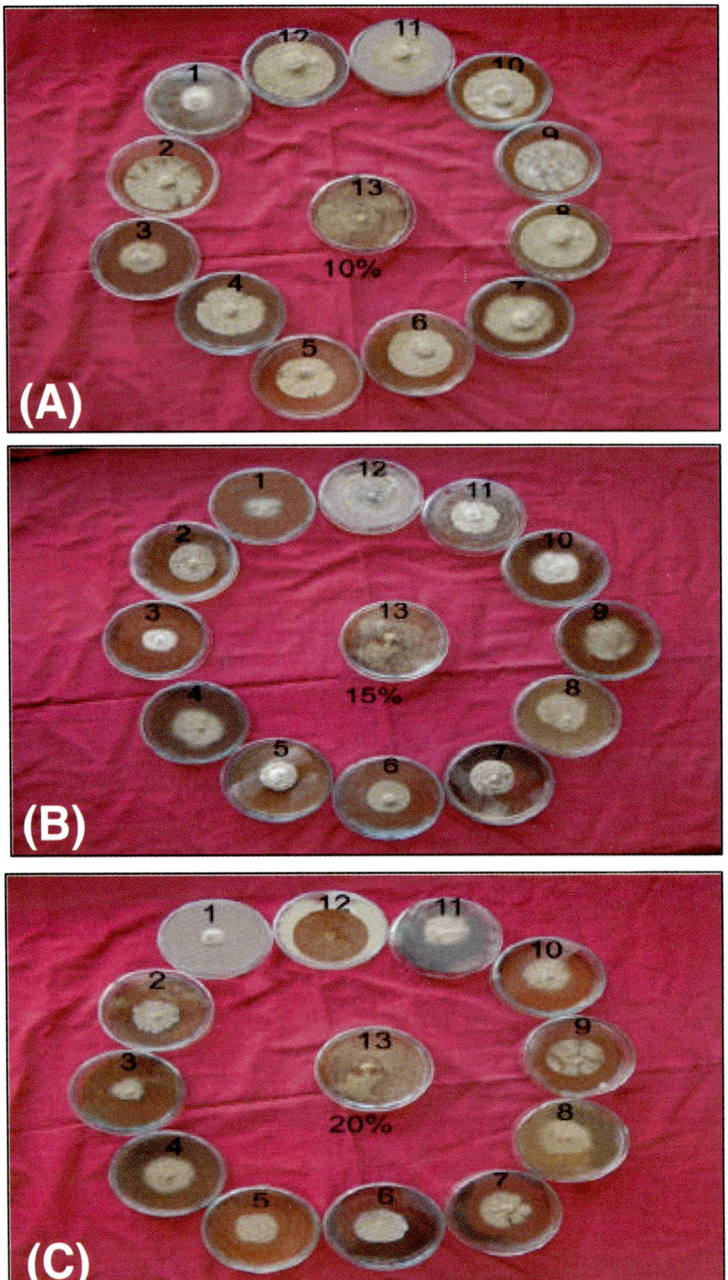

Plate 36.2: *In vitro* Effect of Plant Extracts at 10 per cent (A), 15 per cent (B) and 20 per cent (C) on Mycelial Growth and Inhibition of *S. vesicarium*

1: *A. indica*; 2: *T. erecta*; 3: *O. sanctum;* 4: *E. globulus*; 5: *B. spectabilis*; 6: *L. innermis*; 7: *A. squamosa*; 8: *Z. officinale*; 9: *V. rosea*; 10: *P. hysterophorus*; 11: *I. carnea*; 12: *A. sativum*; 13: Control

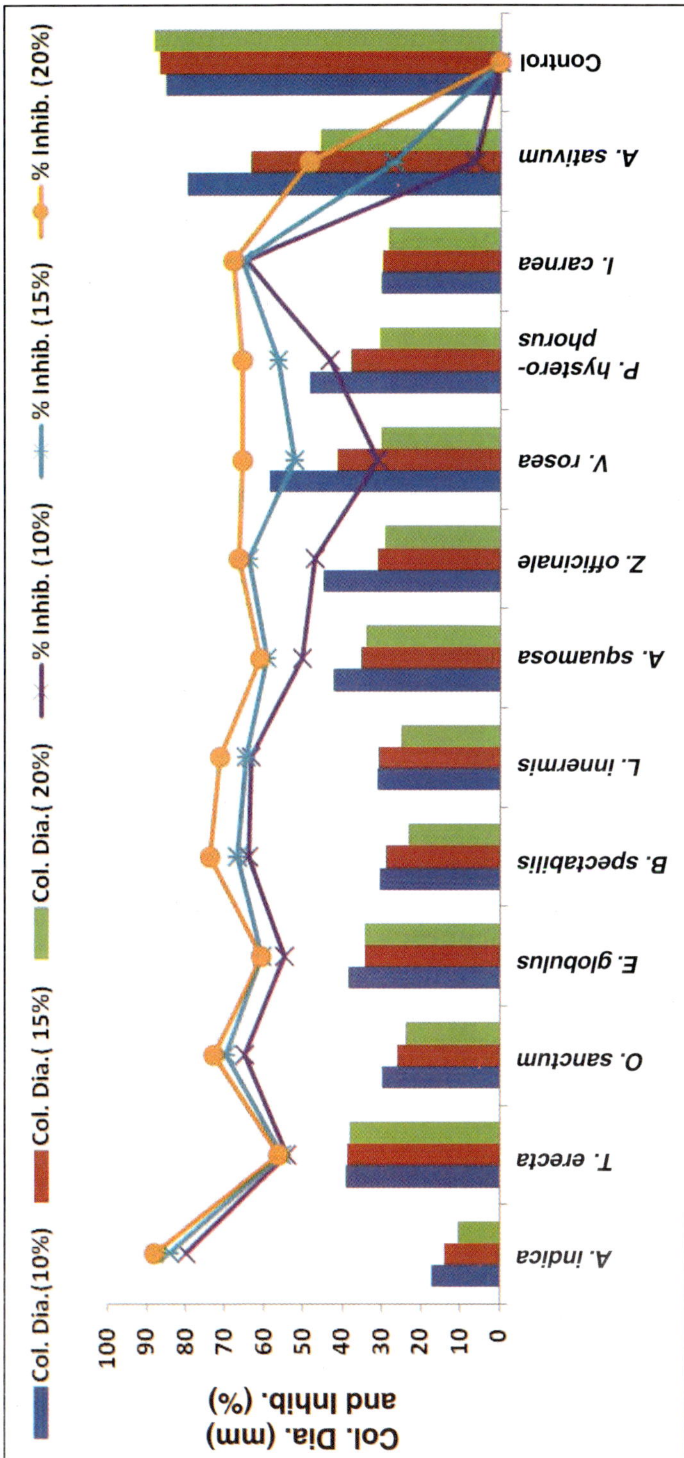

Figure 36.1 *In vitro* Efficacy of Plant Extracts/Botanicals Against Radial Growth and Inhibition of *S. vesicarium*

upon concentrations of the extract used and was decreased in increased concentrations.

Table 36.1: Efficacy of Plant Extracts/Botanicals Against Radial Mycelial Growth and Inhibition of *S. vesicarium*

Treatments	*Colony Diameter (mm)* at Conc.*			*MCD (mm)*	*% Inhibition at Conc.*			*Mean Inhib. (%)*
	10%	*15%*	*20%*		*10%*	*15%*	*20%*	
Neem (*A. indica*)	17.11	13.86	10.58	13.85	79.88 (63.34)	84.00 (66.42)	87.95 (69.68)	83.94 (66.48)
Marigold (*T. erecta*)	39.10	38.73	38.25	38.69	54.04 (47.31)	55.30 (48.04)	56.44 (48.70)	55.26 (48.01)
Tulsi (*O. sanctum*)	29.86	26.16	23.86	26.62	64.90 (53.66)	69.81 (56.67)	72.83 (58.58)	69.04 (56.30)
Eucalyptus (*E.globulus*)	38.50	34.41	34.50	35.80	54.74 (47.71)	60.29 (50.93)	60.71 (51.18)	58.58 (49.94)
Bougainveillia (*B. spectabilis*)	30.56	28.86	23.08	27.50	64.08 (53.17)	66.69 (54.74)	73.72 (59.16)	68.16 (55.64)
Mehandi (*L. innermis*)	31.11	30.78	25.25	29.04	63.43 (52.79)	64.48 (53.41)	71.25 (57.57)	66.38 (54.56)
Custard apple (*A. squamosa*)	42.25	35.40	34.08	37.24	50.34 (45.19)	59.15 (50.27)	61.19 (51.46)	56.89 (48.96)
Ginger (*Z. officinale*)	45.00	31.25	29.41	35.22	47.10 (43.33)	63.93 (53.08)	66.51 (54.64)	59.18 (50.28)
Periwinkle (*V. rosea*)	58.46	41.50	30.25	43.40	31.28 (34.00)	52.11 (46.20)	65.55 (54.05)	49.64 (44.79)
Parthenium (*P.hystero-phorus*)	48.33	37.83	30.41	38.85	43.19 (41.08)	56.34 (48.64)	65.37 (53.95)	54.96 (47.84)
Beshrum (*I. carnea*)	30.20	29.91	28.33	29.48	64.50 (53.42)	65.48 (54.01)	67.74 (55.39)	65.90 (54.27)
Garlic (*A. sativum*)	79.55	63.16	45.41	62.70	6.49 (14.75)	27.11 (21.37)	48.29 (44.02)	27.29 (31.49)
Control	85.08	86.66	87.83	86.52	00.00 (00.00)	00.00 (00.00)	00.00 (00.00)	00.00 (00.00)
S.E. ±	0.18	0.23	0.19	—	0.13	0.22	0.18	—
C.D. (P=0.05)	0.52	0.68	0.57	—	0.38	0.64	0.52	—

*Average of three replications.

Figures in parenthesis are angular transformed values

MCD: Mean Colony Diameter

At 10 per cent concentration, radial mycelial growth of the test pathogen (Plate 36.2A) was ranged from 17.11 mm (*A. indica*) to 79.55 mm (*A. sativum*). However, it was maximum with *A. sativum* (79.55 mm). This was followed by *V. rosea* (58.46 mm),

P. hysterophorus (48.33 mm), *Z. officinale* (45.00 mm), *A. squamosa* (42.25 mm), *T. erecta* (39.10 mm) and *E. globulus* (38.50 mm). Significantly least radial mycelial growth was recorded with *A. indica* (17.11 mm), followed by *O. sanctum* (29.86 mm), *I. carnea* (30.20 mm), *B. spectabilis* (30.56 mm) and *L. innermis* (31.11 mm).

At 15 per cent conc, radial mycelial growth of the test pathogen (Plate 36.2B) was ranged from 13.86 mm (*A. indica*) to 63.16 mm (*A. sativum*). However, maximum radial mycelial growth was recorded with *A. sativum* (63.16 mm), and was followed by *V. rosea* (41.20 mm), *T. erecta* (38.73 mm), *P. hysterophorus* (37.83 mm), *A. squamosa* (35.40 mm), *E. globulus* (34.41 mm) and *Z. officinale* (31.25 mm). Least radial mycelial growth was recorded with *A. indica* (13.86 mm), followed by *O. sanctum* (26.16 mm), *B. spectabilis* (28.86 mm), *I. carnea* (29.91 mm) and *L. innermis* (30.78 mm).

At 20 per cent conc., radial mycelial growth of the test pathogen (Plate 36.2C) was ranged from 10.58 mm (*A. indica*) to 45.41 mm (*A. sativum*). However, highest mean radial mycelial growth was recorded with *A. sativum* (45.41 mm). This was followed by *T. erecta* (38.25 mm), *E. globulus* (34.50 mm), *A. squamosa* (34.08 mm), *P. hysterophorus* (30.41 mm), *V. rosea* (30.25 mm) and *Z. officinale* (29.41 mm). Significantly least mean radial mycelial growth was recorded with *A. indica* (10.58 mm) and this was followed by *B. spectabilis* (23.08 mm), *O. sanctum* (23.86 mm), *L. innermis* (25.25 mm) and *I. carnea* (28.33 mm).

The mean radial mycelial growth/mean colony diameter (MCD) recorded with the plant extracts tested was ranged from 13.85 mm (*A. indica*) to 62.70 mm (*A. sativum*). However, highest mean radial mycelial growth/MCD was recorded with *A. sativum* (62.70 mm). This was followed by *V. rosea* (43.40 mm), *P. hysterophorus* (38.85 mm), *T. erecta* (38.69mm), *A. squamosa* (37.24 mm), *E. globulus* (35.80 mm) and *Z. officinale* (35.22 mm). Significantly least mean radial mycelial growth was recorded with *A. indica* (13.85), and was followed by *O. sanctum* (26.62 mm), *B. spectabilis* (27.50 mm), *L. innermis* (29.04 mm) and *I. carnea* (29.48 mm).

Mycelial Growth Inhibition

Results (Table 36.1 and Figure 36.1) revealed that all the plant extracts tested significantly inhibited mycelial growth of the test fungus over untreated control (00.00 per cent). Further, percentage growth inhibition of the test pathogen was increased with increase in concentrations of the botanicals tested (Plates 36.2A, B and C).

At 10 per cent conc., percentage mycelial growth inhibition (Plate 36.2A) was ranged from 6.49 (*A. sativum*) to 79.88 (*A. indica*). However, significantly highest mycelial growth inhibition was recorded with *A. indica* (79.88 per cent). This was followed by *O. sanctum* (64.90 per cent), *I. carnea* (64.50 per cent), *B. spectabilis* (64.08 per cent), *L. innermis* (63.43 per cent), *E. globulus* (54.74 per cent), *T. erecta* (54.04 per cent) and *A. squamosa* (50.34 per cent). Less than 50 per cent inhibition was recorded with *Z. officinale* (47.10 per cent), *P. hysterophorus* (43.19 per cent), *V. rosea* (31.28 per cent) over untreated control (00.00 per cent). *A. sativum* was found least effective with 6.49 percent inhibition of the test pathogen.

At 15 per cent conc., percentage mycelial growth inhibition (Plate 36.2B) was ranged from 27.11 (*A. sativum*) to 84.00 (*A. indica*). However, significantly highest mycelial growth inhibition was recorded with *A. indica* (84.00 per cent). This was followed by *O. sanctum* (69.81 per cent), *B. spectabilis* (66.69 per cent), *I. carnea* (65.48 per cent), *L. innermis* (64.48 per cent), *Z. officinale* (63.93 per cent), *E. globulus* (60.29 per cent), *A. squamosa* 59.15 per cent), *P. hysterophorus* (56.34 per cent), *T. erecta* (55.30 per cent), *V. rosea* (52.11 per cent). Less than 50 per cent inhibition was recorded with *A. sativum* (27.00 per cent), over untreated control (00.00 per cent).

At 20 per cent conc., percentage mycelial growth inhibition (Plate 36.2C) was ranged from 48.29 (*A. sativum*) to 87.95 (*A. indica*). However, significantly highest mycelial growth inhibition was recorded with *A. indica* (87.95 per cent). This was followed by *B. spectabilis* (73.72 per cent), *O. sanctum* (72.83 per cent), *L. innermis* (71.25 per cent), *I. carnea* (67.74 per cent), *Z. officinale* (66.51 per cent), *V. rosea* (65.55 per cent), *P. hysterophorus* (65.37 per cent), *A. squamosa* (61.19 per cent), *E. globulus* (60.71 per cent) and *T. erecta* (56.44 per cent). Less than 50 per cent inhibition was recorded with Garlic (48.29 per cent), over untreated control (00.00 per cent).

Mean percentage mycelial growth inhibition recorded with all the botanicals tested was ranged from 27.29 (*A. sativum*) to 83.94 (*A. indica*). However, *A. indica* was found most fungi–toxic and recorded significantly highest mean mycelial growth inhibition (83.94 per cent) of the test pathogen. The second and third best plant extracts found were *O. sanctum* (69.04 per cent) and *B. spectabilis* (68.16 per cent). This was followed by *L. innermis* (66.38 per cent), *I. carnea* (65.90 per cent), *Z. officinale* (59.18 per cent), *E. globulus* (58.58 per cent), *A. squamosa* (56.89 per cent), *T. erecta* (55.26 per cent), *P. hysterophorus* (54.96 per cent) and *V. rosea* (49.64 per cent). *A. sativum* (per cent) was found least fungitoxic and recorded minimum mean growth inhibition (27.29 per cent) of the test pathogen.Hence, all the plant extracts tested were found fungi-static/fungi-toxic against *S. vesicarium* and significantly inhibited it's mycelial growth over untreated control.

The fungitoxic/fungistatic effects of the botanicals/plant extract tested might be due to presence of specific antifungal chemical compounds like phenols, tannis, alkaloids, rosinous and non volatile substances.Thus, those plant extracts/botanicals found fungi–toxic/fungi–static against *S. vesicarium* in *in vitro* present study needs to be tested further to confirm their efficacy and if proved effective fungitoxicant could definitely commercialized for economical and eco-friendly management/ control of *Stemphylium* blight of onion as well as other foliage diseases of the crop plants.

Results of the present study on antifungal activity of the botanicals/plant extracts tested against *S. vesicarium*. are in consonance with those reported earlier by several workers. Botanicals Neem (*A. indica*), Tulsi (*O. sanctum*), Mehandi (*L. innermis*), Ginger (*Z. officinale*) and Eucalyptus (*E. globulus*) etc. were reported fungi–static/fungi–toxic against *S. vesicarium* and other *Alternaria* spp. earlier by several workers (Prasad and Barnwal, 2004; Tiwari and Srivastava, 2004; Kumari *et al.*, 2006; Bhatiya and Awasthi, 2007; Verma *et al*, 2007).

Thus, as plant species/botanicals with antifungal/fungitoxic potentials and plant based byproducts have been proved to be non- phyto toxic, biodegradable and eco–friendly; they needs to be exploited commercially and incorporated in the integrated management of the crop diseases, which is of prime importance for conservation of environment and sustainable agriculture.

References

Amaresh, Y.S. and Nargund, V.B. (2004). *In vitro* evaluation of fungicides against *Alternaria helianthi*, causing leaf blight of sunflower. *Ind. J. Pl. Pathol.*, **22(1&2)**: 79-82.

Barnwal, M.K., Prasad, S.M. and Mait, D. (2003). Efficacy of fungicides and bioagents against *Stemphylium* blight of onion *Indian Phytopath.*, **56(3)**: 291-292.

Bhatiya, B.B. and Awasthi, R.P. (2007). *In vitro* evaluation of some antifungal plant extracts against *Alternaria brassicae*, causing *Alternaria* blight of Rapeseed-Mustard. *J. Pl. Dis. Sci.*, **2 (2)**: 126-131.

Gupta, R.P. (1986). *Stemphylium* blight causes total failure of seed crop in North India. *Allium Improvement News Letter*, **5**: 22-24.

Gupta, R.P. and Srivastava, P.K. (1988). Control of *Stemphylium* blight of onion bulb crop. *Indian Phytopath.*, **41 (3)**: 495-496.

Kharakwal, V.P., Patni, C.S. and Adhikari, R.S. (2006). Effect of some fungicides on spore germination and growth of *Alternaria solani in vitro*. *J. Pl. Dis. Sci.*, **1 (1)**: 128-130.

Khosla, K., Thakur, B.S. and Bharadwaj S.S. (2007). Chemical management of *Stemphylium* blight of garlic. *Pl. Dis. Res.*, **22 (1)**: 47-51.

Kumari, L., Shekhawat, K.S. and Rai, P.K. (2006). Efficacy of fungicides and plant extracts against *Alternaria* blight of Periwinkle (*Catharanthus roseus*).

Patni, C.S., Kolte, S.J. and Awasthi, R.P. (2005). Efficacy of botanicals against *Alternaria* blight (*Alternaria brassicae*) of mustard. *Indian Phytopath.*, **58 (4)**: 426-430.

Prasad, S.M. and Barnwal, M.K. (2004). Evaluation of plant extracts in management of *Stemphylium* blight of onion. *Indian Phytopath.*, **57 (1)**: 110-111.

Nene, Y.L. and Thapliyal, P.N. (1993). Evaluation of fungicides. In: *Fungicides in Plant Disease Control*, 3rd Edn. Oxfrod, IBH Publishing Co., New Delhi, pp. 331.

Shahanaz, E., Razdan, V.K. and Raina, P.K. (2007). Survival, dispersal and management of foliar blight pathogen of onion. *J. Mycol. Pl. Pathol.*, **37**: 210-214.

Tiwari, B.K. and Srivastava, K.S. (2004). Studies on bio-efficacy of some plant extract against pathogens of onion. *News Letter NHRDF*, **24(1)**: 6-1.

Verma, K.P., Singh, S. and Gandhi, S.K. (2007). Variability among *Alternaria solani* isolates causing early blight of tomato. *Indian Phytopath.*, **60(2)**: 180-186.

Vincent, J.M. (1927). Distortion of fungal hyphae in the presence of certain inhibitors. *Nature*, pp. 159-180.

Chapter 37

Effect of Different Dates of Sowing on Stigma Receptivity in Parental Lines of Sorghum Hybrids

N.B. Mehetre[1], S.T. Rathod[1] and B.G. Kamble[1]

ABSTRACT

The investigation was carried out to study the "Effect of Different Dates of Sowing on Stigma Receptivity in Parental Lines of Sorghum Hybrids". The experiment was laid out in the field, Department of Botany, Dr. Panjabrao Deshmukh Krishi Vidyapeeth, Akola during *kharif* 2003-2004.The result indicated that in D_1, D_2 and D_3 sowing dates, stigma of seed parent MS-27 A remains receptive up to 6 days and highest receptivity on 0 days and lowest on 6 days of starvation period whereas stigma receptivity of other seed parent AKMS-14A, MS-70A and MS-296A remains receptive up to 5 days of starvation as judged by seed setting percentage.

Keywords: Different sowing dates, Hybrids, RBD, Sorghum.

Introduction

Sorghum [*Sorghum bicolor* (L.) Moench] is the 5th major cereal crop in the agricultural scenario of the globe. The exploitation of hybrid vigour has increased the

---

1 Department of Agricultural Botany, Dr. Panjabrao Deshmukh Krishi Vidyapeeth, Akola, M.S., India

productivity of this crop by three fold than the yield from the traditional varieties. It is a valuable food crop for grain to mankind and fodder to cattle.

There are many reasons for low yield of released hybrids and one of the reason is the improper nicking of male sterile line and restorers of hybrids (Chopde, 1973). Hence, improved seed production technology will help to improve the present low level of seed yield and spread of hybrids.

Stigma Receptive

For better seed production of hybrids, it is essential that, the male sterile (A) line and restorer (R) line should synchronize in flowering. Delayed or early flowering of any parent adversely affect the seed setting (Singh *et al.*, 1985). Among the released hybrids serious synchronization problem is often observed with the hybrid seed production of CSH-5 and CSH-9 (Singh and Nayeem, 1980).

The stigma receptivity studies of male sterile lines will help in predicting proper planting time of male and female lines for maximum seed set (Shelar and Patil, 1993). Further, it will help to find out the male sterile lines with longer stigma receptivity period. Stigma receptivity studies give valuable information as regards adaptability of females in different environment.

Materials and Methods

The investigation was carried out to study the "Effect of Different Dates of Sowing on Stigma Receptivity in Parental Lines of Sorghum Hybrids". The experiment was laid out in the field, Department of Botany, Dr. Panjabrao Deshmukh Krishi Vidyapeeth, Akola during *kharif* 2003-2004.

In the present investigation three planting dates and eight genotypes were studied in Randomizer Block Design (Factorial) with three replications, plot size 4m x 2.7 m, spacing row to row 45 cm and plant to plant 15 cm in three different sowing dates, D_1 25th June 2003, D_2 1st July 2003 and D_3 7th July 2003.

Results and Discussion

In D_1 sowing date, parent MS-27A exhibited highest stigma receptivity (93.21 per cent) followed by MS-70A (90.83 per cent), AKMS-14A (88.21 per cent) and MS-296A (84.05 per cent) at '0' days starvation period. All the parents in D1 was receptive up to '4' days starvation period except MS-27A was receptive up to '5' days starvation period. In general AKMS-14A, MS-70A and MS-296A had better seed setting percentage upto'2' days of starvation period after complete flowering where as in MS-27A had better seed setting percentage upto'3' days of starvation period in D_1 sowing date. Stigma receptivity of MS-27A exhibited highest (53.36 per cent) and MS-296A exhibited lowest (41.63 per cent) whereas AKMS-14A exhibited (42.86 per cent) and MS-70A (44.19 per cent), calculated from the mean performance (Table 37.1).

In D_2 sowing date, parent AKMS-14A exhibited lowest seed setting (85.19 per cent) followed by MS-296A (87.43 per cent), MS-70A (87.56 per cent) and MS-27A (89.24 per cent) at '0' days starvation period. All the parents in D_2 was receptive up to '4' days starvation period except MS-27A was receptive up to 5 days starvation period. In general AKMS-14A, MS-70A and MS-296A had better seed setting percentage up to 2 days of starvation period after complete flowering (Table 37.2).

Table 37.1: Performance of Male Sterile 'A' Lines for Stigma Receptivity Studies in D_1 Sowing Date

Genotypes (F2)	*Starvation Period in Days (F_1)*						
	0	*1*	*2*	*3*	*4*	*5*	*6*
AKMS-14A	88.21	78.26	60.26	53.25	20.10	0.00	0.00
MS-27A	93.21	84.24	76.01	64.79	33.41	21.90	0.00
MS-70A	90.83	88.13	69.87	49.95	15.55	0.00	0.00
MS-296A	84.05	75.20	61.91	45.02	25.25	0.00	0.00
Mean	89.07	80.20	67.01	53.25	23.57	5.47	0.00

	'F' Test	*SE (m)*	*CD at 5%*
Sowing Dates (F_1)	Sig.	0.77	1.55
Genotypes (F_2)	Sig.	0.58	1.17
Interaction (F_1 x F_2)	Sig.	1.55	3.11

Table 37.2: Performance of Male Sterile 'A' Lines for Stigma Receptivity Studies in D_2 Sowing Date

Genotypes (F2)	*Starvation Period in Days (F_1)*						
	0	*1*	*2*	*3*	*4*	*5*	*6*
AKMS-14A	85.92	77.30	56.22	48.18	18.34	0.00	0.00
MS-27A	89.24	81.31	68.12	58.28	28.39	18.56	0.00
MS-70A	87.56	75.10	65.75	42.16	12.48	0.00	0.00
MS-296A	87.43	76.17	67.72	40.53	20.41	0.00	0.00
Mean	87.53	77.47	64.45	47.28	19.90	4.64	0.00

	'F' Test	*SE (m)*	*CD at 5%*
Sowing Dates (F_1)	Sig.	0.62	1.25
Genotypes (F_2)	Sig.	0.47	0.94
Interaction (F_1 x F_2)	Sig.	1.24	2.50

In D_3 sowing date, parent AKMS-14A exhibited highest seed setting (86.21 per cent) followed by MS-70A (85.34 per cent), MS-27A (84.61 per cent) and MS-296A (81.27 per cent) at '0' day starvation period. All the parents AKMS-14A, MS-70A and MS-296A was receptive up to 4 days of starvation except MS-27A which was receptive up to 5 days of starvation showed (15.19 per cent) seed setting (Ross, 1957).

In general AKMS-14A, MS-27A, MS-70A and MS-296A had better seed setting up to 2 days starvation and subsequently reduced up to 6 days of starvation.

Table 37.3: Performance of Male Sterile 'A' Lines for Stigma Receptivity Studies in D_3 Sowing Date

Genotypes (F2)	*Starvation Period in Days (F_1)*						
	0	*1*	*2*	*3*	*4*	*5*	*6*
AKMS-14A	86.21	80.45	60.13	45.28	15.27	0.00	0.00
MS-27A	84.61	78.00	64.18	54.46	27.26	15.19	0.00
MS-70A	85.34	71.15	66.78	41.41	10.36	0.00	0.00
MS-296A	81.27	75.88	58.19	38.33	17.27	3.79	0.00
Mean	84.35	76.37	62.32	44.87	17.54	3.79	0.00

	'F' Test	*SE (m)*	*CD at 5%*
Sowing Dates (F_1)	Sig.	0.80	1.60
Genotypes (F_2)	Sig.	0.60	1.21
Interaction (F_1 x F_2)	Sig.	1.60	3.20

From the data, it was observed that seed setting percentage was significantly better up to 3 days of starvation period after complete flowering in all sowing dates.

I[st] sowing date exhibited early flowering for all the genotypes studied where 2nd and 3rd sowing dates didn't make any significant deviation. Flowering was delayed in 2nd sowing dates. Duration of flowering was 4 to 5 days in 1st, 4.33 to 6 days in 2nd and 4 to 6 days and in 3rd sowing date.

Stigma receptivity was studied for male sterile (A) lines *viz.* AKMS-14A, MS-27A, MS-70A and MS-296A of CSH-14, CSH-16, SPH-840 and SPH-388 respectively. On an average, stigma receptivity was significantly highest at '0' days starvation (Shellar and Patil, 1993) and subsequently reduced due to delay in pollination. Seed setting percentage in seed parents was significantly better up to 2 and 4 days of starvation after complete flowering (Patil *et al.*, 1998).

These results indicated that stigma will be receptive up to '3' days of starvation or '7' days from start of flowering in *kharif* season. Stigma receptivity of seed parent MS-104 A and 27 A of hybrid CSH-15R and CSH-16 was significantly superior at '0' day starvation but satisfactory up to '2' days in *kharif* and rabi at Parbhani and Hyderabad respectively (Anonymous, 1998-1999).

References

Anonymous (1998-1999). Annual Report of Research work submitted to food crop science submitted by Seed Technology Research Unit (N.S.P.) Hyderabad, Akola, Parbhani and Rahuri pp. 71-78.

Chopde, P.R. (1973). Nicking studies in hybrid seed production of CSH-2 and CSH-3. *Sorghum Newsletter*. **16**: 55-56.

Patil, R.B., Suryawanshi, Y.B., Shinde, P.Y. and Shelar, V.R. (1998). Studies on stigma receptivity and pollen viability in parental lines of sorghum hybrids. *Seed Tech News.*, **28 (4)**: 71.

Ross, W.M. (1957). Stigma receptivity in cytoplasmic male sterile. *Sorghum Agron. J.*, **9(4)**: 219-220.

Shelar, V.R. and Patil, R.B. (1993). Synchronization studies in *kharif* sorghum hybrids. *Seed Tech. News*, **21 (2)**: 86-88.

Singh and Nayeem (1980). Parental stability for flowering behavior in relation to seed production in sorghum. *Indian J. Agric. Sci.*, **50**: 202-207.

Singh, A.R., Nayeem, K.A. and Chopde, P.R. (1985). Stigma receptivity studies in cytoplasmic male sterile lines of sorghum hybrids. *Seed Res.*, **7(2)**: 92-97.

Chapter 38

Incidence of Arbuscular Mycorrhizal Association in *Striga gesnerioides* (Willd) Vatke. Oester, var *gesnerioides*: Root Parasitic Medicinal Plant and its Host *Lepidagathis cuspidata* Nees from Maharashtra, India

Vishal R. Kamble[1] and Dinesh G. Agre[2]*

ABSTRACT

Occurrence of mycorrhizal fungi is universal. Symbiotic association of AM is found in most of the families of angiosperms, gymnosperms, pteridophytes and bryophytes. However, the incidence of mycotrophy in parasitic plant species is rarely documented. *Striga gesnerioides* (Willd) Vatke. Oester, Var *gesnerioides*

1 Department of Botany, Bhavan's College Andheri (West), Mumbai – 400 058, M.S., India

2 Department of Biology, Utkarsha Vidyalaya and Jr. College, Virar (W) – 401 303, M.S., India

* Corresponding Author E-mail: vrksiddhant@rediffmail.com

(Scrophulariaceae) is common parasite on *Lepidagathis cuspidata* Nees (Acanthaceae) roots at high altitudes in open areas throughout Maharashtra and grows in very poor on hilltops with gravelly soil. *Striga gesnerioides* has medicinal properties as it used against diabetes. In present investigation *S. gesnerioides* and its host *L. cuspidata* collected from Panhala Hill of Kolhapur District located in Western Ghats of Maharashtra were found to be mycorrhizal. The extent of root colonization in the parasite and host was found to be 55.33 per cent and 30.76 per cent respectively. AM fungal spores of two genera *viz., Glomus and Gigaspora* were isolated from rhizosphere soil, which showed great diversity at species level.

Keywords: *Arbuscular mycorrhizal fungi, Medicinal plants, Root colonization.*

Introduction

Striga is commonly known as witchweed (Scrophulariaceae) although they are obligate parasites; do not obtain all their nutrients from their host roots. Different species of *Striga* parasitize many important plants such as corn, cereals, sugarcane and tobacco in Africa, Australia, India and Indonesia. In India it is a well known semi root parasite of cereals, maize and millets (Mehrotra, 1980), sugarcane (Luthra, 1921), and *S. euphrasioides* is known to be parasitic on paddy (Rao *et al.*, 1953).

Five species of *Striga* are known to flora of Maharashtra (Godbole and Prasad, 2001). Four species of *Striga viz., S. angustifolia* (D. Don) Sald., *S. asiatica* (L.) O. Ktze., *S. densiflora* (Benth.) Benth., and *S. gesnerioides* (Willd.) Vatke Oester, Var *gesnerioides* are distributed at Amba, Ajara, Dajipur and Gaganbavda of Kolhapur district (Yadav and Sardesai, 2002). However, in present report *S. gesnerioides* was found growing in poor gravelly soil of hill-slope on host plant *L. cuspidata* Nees (Acanthaceae) roots in Panhala hills. The study area is situated at an altitude of 3100 ft and 18 km northwest of Kolhapur, in Maharashtra state. Panhala is located at 16°49′N 74°07′E/16.82°N 74.12°E/16.82; 74.12.

Although *Striga* has been reported as parasitic plant, however some species exhibit certain medicinal potential. *Striga asiatica* is used in poisonous bites, dental disorders and as an appetizer; Leaves of *S. densiflora* are useful in haematuria (Chetty, *et al.*, 2008); Earlier, Rao, (1990) and Hiremath *et al.* (1990), showed significant antifertility efficacy of *S. lutea* in mice; Choudhury, *et al.* (1998) showed that, flowers of *S. senegalensis* has anti-implantation activity in female albino rats. Okpako and Ajaiyeoba, (2004), reported antimalarial activity in *S. hermonthicat*. Harisha *et al.* (2001), reported antihistaminic and mast cell stabilizing potential in *S. orobanchioides*. Evidences suggest that most of the species have been tested on experimental animals. Ethnobotanical data on *S. gesnerioides* reveals that fruits (Ambasta, 1986; Jain, 1996; Sinha and Sinha, 2001) and whole plant is considered as medicinal since it is used against diabetes (Chetty, *et al.*, 2008).

Since most of the *Striga* species (exceptionally *S. gesnerioides*) are found parasitically associated with roots of agronomical crops and hence most of the time farmers eradicate them with selective weedicide sprays. As eradication of parasite is the only means to protect crop, the huge quantity of medicinal biomass is inadvertently destroyed. *Striga* reproduces only by means of very minute seeds. The germination occurs only after stimulus provided by the root exudates of specific host plant which enables the seeds to develop. The root of the parasite form haustoria which penetrate the host roots, drawing nourishment from the host. *S. gesnerioides*, unlike other species is common root parasite on *Lepidagathis* roots at high altitudes in open areas throughout Maharashtra and it is not associated with agriculture crops. That is why to obtain huge biomass of whole plant against diabetes, its cultivation on host roots–*Lepidagathis*, which is non economical plant in present context will be appreciable. To introduce a cultivation technique for wild plant species it is very important to study natural association of microorganisms if any with plant along with environmental conditions. The present chapter is an attempt to explore AM fungi associated with host *Lepidagathis* and parasitic medicinal plant *S. gesnerioides*.

Arbuscular mycorrhizal (AM) fungi normally colonizes almost all tropical plants, however the incidence of mycotrophy in medicinal plants is less documented compared to studies on forestry and crop species (Bagyaraj, 1995). AM fungi are ubiquitous in distribution in 80-90 per cent of the plants ranging from Bryophytes to flowering plants. They are regularly formed by vascular plants from wide range of habitat. AM fungi have played a key role in plant evolution on earth as well as on the development and maintenance of the structure and diversity of terrestrial ecosystems (Palenzuela, *et al.*, 2008). Most of the plants depend on mycorrhizas to thrive particularly in fragile and stressed environment.

There is steady increase in the cultivation of medicinal plants to maintain a steady supply to support the increasing demand with the rise in the consumption of herbal products in India. Since last few decades the use of herbal to ameliorate human sufferings are gaining importance at par with the synthetic drugs which has inspired the Government of India and plant scientists to promote the conservation of natural flora including medicinal plants and their propagation. This is essentially required to meet the increasing demand of herbals (Roy *et al.*, 2007). Due to the importance of AM symbiosis for plant establishment and development in stress condition, a study was initiated to determine the AM colonization and diversity in *S. gesnerioides*–medicinal anti-diabetic parasitic plant.

Materials and Methods

The roots and rhizospher soil samples of host-parasite association were collected from study area during Nov.-Dec. 2010. At least three associations of plants were screened for mycorrhizal colonization. Host (*L. cuspidata*) and parasite (*S. gesnerioides*) root connections were traced out carefully. Host roots, parasite roots and host-parasite root junctions were carefully selected for study of AM–fungi colonization study. The non-pigmented roots were cleaned and stained following the method of Phillips and Hayman (1970), while the pigmented roots were treated following the method of Kormanik *et al.* (1980). One hundred root segments (1 cm) were randomly selected for microscopic observation and quantification of mycorrhiza was calculated in percentage (Read *et al.*, 1976). Mycorrhizal spores and sporocarps were recovered from soil by wet sieving and decanting method (Gerdemann and Nicolson, 1963).

The spores and sporocarps were identified following the synaptic keys of Gerdmann and Trappe (1974), Hall and Fish (1979) and Trappe (1982). A species was considered mycorrhizal if the root samples showed hyphae, vesicles or arbuscular colonization (Pendleton and Smith, 1983; Pond *et al.*, 1984).

Results and Discussion

The results confirm AM association in roots of *S. gesnerioides*, and *L. cuspidate*. Percent root colonization in both host and parasite was less than colonization in root junction region (Table 38.1). Arbuscules colonization was excellent in *S. gesnerioides* roots. However, vesicles were found comparatively smaller in size than in host *L. cuspidata* and *L. cuspidata–S. gesnerioides* root junction region. Variation in size and shapes of vesicles as well as in percent colonization was observed. An explanation for this variation might be that, the present biological association of host and parasitic plant is a kind of natural biological stress condition in which AM fungus has tried to establish colonization. The present study confirms the AM association in *Striga*. Colonization of AM fungi was indicated by the presence of darkly stained vesicles and arbuscules in the roots. The present study confirms the earlier reports that, Scrophulariaceae members of diverse habitat have showed variation in AM colonization but are rare in *Striga*. Earlier work on AM colonization in Scrophulariaceae have showed that river bank plants *viz., Lindernia antipoda* (L.) Alston, *L. hysopiodes* Haines, and *Verbascum chinense* (L.) Santapau etc., were non mycorrhizal; whereas, amongst the aquatic-semi aquatic plants *Limnophila indica* (L.) Druce was found to be mycorrhizal in association and *Peplidium maritimum* (L.) Aschers showed absence of AM while *S. angustifolia* has been reported mycorrhizal from Costal dry evergreen forest of Thanjavur, (Ragupathy, *et al.*, 1988).

Table 38.1: Colonization Percentage of AM Fungi in the Roots of *S. gesnerioides* and *L. cuspidate*

Sl.No.	Plant Species	% Colonization
1.	*L. cuspidata* (Host)	30.76 per cent
2.	*S. gesnerioides* (Parasite)	53.33 per cent
3.	*S. gesnerioides–L. cuspidata* (Root junction- region)	83.87 per cent

The diversity of fungal spore was quite good. AM spore count recorded 590-600 spores per 10g of soil sample. AM species recorded during thet study included *Enterophospora* sp., *Gigaspora* sp., *Glomus epigaeum, G. macrocarpum, G. occulatum*, etc. Amongst these only *G. macrocarpum* and *Gigaspora* (showed diversity at species level) were found in abundance. Besides, other species of AM fungi isolated in less numbers and included *Archaeospora sp., Glomus coremoides, G. aggregatum, G. ambisporum, G. fasciculatum, etc.*

This study confirms the fact that AM fungi are an important part of host-parasite association in plants. Since the role of AM fungi is still not fully understood in such an association, it needed to give more focus on studies on similar plants. Before introducing any cultivation technique for wild medicinal plant species it is very important to study natural association of microorganisms like AM fungi. The results from this study imply that AM associations might be functional and significant component of host and parasitic plants.

Acknowledgements

Thanks are to Dr. R. M. Mulani, S. R. T. M. University, Nanded for his valuable suggestions. Thanks are also to due to Principal Dr. V. I. Katchi, Dr. M. S.Chemburkar and Prof. Sriharsha Head of Botany Department of Bhavan's College, for providing laboratory facilities.

References

Ambasta, S.P. (Ed.) (1986). *The Useful Plants of India*. CSIR, New Delhi.

Bagyaraj, D.J. (1995). Mycorrhizal association in crop plants and their utilization in agriculture. In: *Benificial Fungi and their Utilization,* (Eds.) M.C. Nair and Bala Krishnan. Scientific publisher, Jodhpur, India, pp. 61-71.

Chetty, K.M., Sivaji, K. and Rao, K.T. (2008). *Flowering Plants of Chittoor District Andhra Pradesh India*. Student Offset Printers, Tirupati, India, pp. 239.

Choudhury, M.K., Sani, U.M. and Mustapha, A. (1998). Antifertility activity of the flowers of *Striga senegalensis* (Scrophulariaceae) Article first published online: 18 DEC Phytotherapy Research John Wiley and Sons, Ltd.

Gerdemann, J.W. and Trappe, J.M. (1974). *Mycologia. Memoir,* **5**: 1-76.

Gerdemann, J.W. and Nicolson, T.H. (1963). Spores of mycorrhizal endogone species extracted from soil by wet sieving and decanting. *Trans. Br. Mycol. Soc.,* **46**: 235-244.

Godbole and Prasad (2001). In Singh *et al.* (Eds.), *Flora of Maharashtra State Dist.* (2) 550-551.

Hall, I.R. and Fish, B.J. (1979). A key to the Endogonaceae. *Trans. Br. Mycol. Soc.,* **73**: 261–270.

Harisha, M.S., Nagura, M. and Badami, S. (2001). Antihistaminic and mast cell stabilizing activity of *Striga orobanchioides. Journal of Ethnopharmacology,* **76**: 2, 197-200.

Hiremath, S.P. *et al.* (1990). Antifertility Activity of *Striga lutea*-Part I. *Indian Journal Physiol. Pharmacol.,* **34**: **1**, 23-25.

Jain, S.K. (ed.) (1996). *Ethnobotany in Human Welfare*. Deep Publications, New Delhi.

Kormanik, P.P., Bryan, W.C. and Shultz, R.C. (1980). Procedure and equipment for staining large number of plant root samples for endomycorrhiza. *Can. J. Microbiol.,* **26**: 536-538.

Luthra, J.C. (1921). Striga as root parasite of sugarcane. *Agr. J. India,* **16**: 591-523.

Mehrotra, R.S. (1980). *Plant Pathology*. Tata McGraus-Hill Pub., India, pp. 740-748.

Okpako, L.C. and Ajaiyeoba, E.O. (2004). *In vitro* and *in vivo* antimalarial studies of *Striga hermonthica* and *Tapinanthus sessilifolius* extracts. *Afr. J. Med. Sci.*, **33**: **1**, 73-5.

Palenzuela, J. Ruiz-Girela, M., Barea, J. M. and Azcon-Aguilar, (2008). Diversity of Arbuscular Mycorrhizal fungi in Rhizosphere of endangered and/or endemic plants in Sierra Nevada (Granada, Spain) National Park. In: (Abstracts Book), An International Conference on Plant-Microbial Interactions, Krakow, Poland, 40.

Pendleton, R.L. and Smith (1983). *Oecologia*, **59**: 296-301.

Phillips, J.M. and Hayman, D.S. (1970) Improved procedure for clearing roots and staining parasitic and vesicular arbuscular mycorrhizal fungi for rapid assessment of infection. *Trans. Br. Mycol. Soc.*, **55**: 158–161.

Pond, E.C., Menge, J.A. and Jarrel, W.M. (1984). *Mycologia*, **76**: 74-84.

Ragupathy, V. Mohankumar and Mahadevan, A. (1988). Distribution of VAM in Thanjavur district flora. In: *Proceedings: Mycorrhiza for Green Asia*, First Asian Conference on Mycorrhizae, (Eds.) A.N. Mahadevan, Raman and K. Natarajan. CAS in Botany University of Madras, India, p. 95-98.

Rao, G.K. (1990), Antifertility efficacy of the plant *Striga lutea* (Scrophulariacae) on rats. *Contraception*. **42**: 4, 467-77.

Rao, C.D., Sinha, A.K. and Gupta, B.G. (1953). A note on Striga euphrasioides on paddy and its control, *Madras Agr. J.*, **40**: 381-382.

Read, D.J., Kouchki, H.K. and Hodson, J. (1976). Vesicular arbuscular mycorrhiza in natural vegetation system. The occurrence of infection. *New Phytol.*, **77**: 641-653.

Roy A.K., Singh, A.N. and Gautam, N.K. (2007). Rhizosphere AM fungi of some rare medicinal plants. In: *Rhizosphere Biotechnology: Plant Growth Retrospect and Prospect*, (Eds.) A.K. Roy, B.N. Chakraborty, D.S. Mukadam and Rashmi. Scientific Publishers India, pp. 201-206.

Sinha, P.K. and Sinha, S. (2001). Ethnobotanical role of indigenous and ethnic societies. In: *Biodiversity Conservation, Human Health Protection and Sustainable Development*. Surbhi Publication, Jaipur.

Trappe, J.M. (1982). Synoptic keys to the genera and species of Zygomycetes mycorrhizal fungi. *Phytopathology*, **72**: 1102-1108.

Yadav, S.R. and Sardesai, M. (2002). *Flora of Kolhapur District*, p. 334.

Chapter 39

Effect of Passage on the Development of Carbendazim Resistance in *Alternaria alternata* L. Causing Fruit Rot of Pomegranate

S.S. Bharade[1], S.S. Kamble[2] and D.K. Kirwale[3]*

ABSTRACT

Pomegranate (*Punica granatum* L.) is one of the commercially important fruit in India. Fruit rot of its causes yield losses, in the present research work it is shown that instead of spraying only one fungicide use of carbendazim alternately

1 Mahyco Research Foundation Trust, Department of Botany, Badrinarayan Barwale Mahavidyalaya, Jalna, M.S., India

2 Mycology and Plant Pathology Laboratory, Department of Botany, Shivaji University, Kolhapur, M.S., India

3 Rayat Shiksan Sansthas, Mahatma Phule Science and Commerce College, Panvel, Navi Mumbai, M.S., India

* Corresponding Author E-mail: sunita.bharade29@gmail.com

with Diathane-M-45, Kavach, and captafol @ 500 µg/ml inhibited the growth of pathogen at 4th and 3rd passage respectively. Similarly, carbendazim when used in mixture with carbendazim + Diathane-M-45, carbendazim + Kavach, carbendazim + captafol growth of the pathogen was completely inhibited at 2nd passage only.

Keywords: *Fungicides, Fruit rot, Passage, Alternaria alternata (Fr.)*

Introduction

Alternaria alternata (Fr.) Keissler causes fruit rot of pomegranate (*Punica granatum* L.). Pomegranate is one of the commercially important fruit in our country. In Maharashtra many farmers growing pomegranate. But recently due to change in climatic conditions such as delayed rains, eneven temperatures, inadequate storage and transit the fruit is attacked by many fungal diseases which causes economical losses to the growers. In the present research work it is investigated that fungicides instead of used alone use alternately and in combination with other fungicides to minimize the fruit rot.

Materials and Methods

After determination of MIC (Minimum inhibitory concentration) the effect of continuous and alternate treatments of fungicides with two different modes of actions and a mixture of both on the development of resistance in sensitive isolates Aa-2 (500 µg/ml) MIC was studied by food poisoning technique (Dekker and Gielink, 1979), *in vitro*. The fungicides used were chlorothalonil (Kavach 75wp), captafol, Diathane-M-45. These fungicides were tested at 500ppm.Aa-2 in each passage was cultured on plates with carbendazim (Bavistin 50WP), (500 µg/ml). Plates without fungicides served as control. A 8 mm agar disc of freshly grown culture taken from the culture of previous passage of the same isolate was placed at the centre of each plate in triplicate. *In vivo* studies were carried out on pomegranate fruit, fruits were surface sterilized with alcohol and with the help of 8mm cork borer 15mm deep well was prepared and filled with fungicidal solutions 24hrs before inoculation of the pathogen. Mycelial suspension was placed in it, closed with pomegranate disc and wrapped with paper towels and incubated in dark at room temperature 26 ± 3°C for 12 days. Percent growth inhibition was calculated up to 7th passage.

Results and Discussions

The Aa-2 isolate on carbendazim continuously for 7 successive passages significantly increases the resistance. Use of carbendazim alternately with Diathane-M-45, Kavach, and captafol inhibited the growth of pathogen at 4th, 4th. 3rd passage respectively (Table 39.1). Similarly, carbendazim when used in mixture with carbendazim + Diathane-M-45, carbendazim + Kavach, carbendazim + captafol growth of the pathogen was completely inhibited at 2nd, passage (Table 39.2). According to Griffin (1981) the alternately used fungicide must have different mode of action. Multisite action of carbendazim might responsible for the development of resistance in fruit rot of pomegranate. alternate ly use of ediphensophos in reducing carbendazim resistance in *Septoria nodorum* and *Cercosporella herptotrichoide* in cereals have been shown by Horsten (1979). Hartill (1983) reported that alternate use of Maneb to metalaxyl to control late blight of potato, Combination of benomyl with captan reduces the rate of infection of *Venturia inaequalis* (Shabi and Glpatric, 1981). Gangawane *et al.* (1983) determined the development of metalaxyl resistance in *Phytopthora infestans*. Recently, Bhale and Gogle (2008) reported the development of carbendazim resistance in *Alternaria spinaciae.*

Table 39.1: Effect of Exposure of Aa_2 to Carbendazim Continuous and Alternating with Other Fungicides on the Development of Resistance During 7th Successive Passages *in vitro*

Fungicides	*Percentage Growth Inhibition and Passage Numbers*						
500ppm	*1*	*2*	*3*	*4*	*5*	*6*	*7*
Carbendazim individual	9.33	10.66	11.66	12.66	13.33	14.33	20.00
Carbendazim alters Diathane-M-45	9.33	11.33	14.33	11.66	00.00	00.00	00.00
Carbendazim alters Kavach	9.33	13.66	12.66	00.00	00.00	00.00	00.00
Carbendazim alters captafol	9.33	10.00	00.00	9.00	00.00	00.00	00.00
S.E.	00.00	0.69	2.82	0.90	2.74	2.50	8.66
C.D.	00.00	3.17	6.45	12.9	7.61	7.61	11.08

Table 39.2: Effect of Exposure of Aa_2 to Carbendazim Continuous and Mixture with Other Fungicides on the Development of Resistance During 7th Successive Passages *in vitro*

Fungicides	*Percentage Growth Inhibition and Passage Numbers*						
500ppm	*1*	*2*	*3*	*4*	*5*	*6*	*7*
Carbendazim individual	9.33	10.66	11.66	12.66	13.33	14.33	20.00
Carbendazim+ Diathane-M-45	00.00	00.00	00.00	00.00	00.00	00.00	00.00
Carbendazim+ Kavach	00.00	00.00	00.00	00.00	00.00	00.00	00.00
Carbendazim+ captafol	00.00	00.00	00.00	00.00	00.00	00.00	00.00
S.E.	00.17	2.00	2.19	2.37	2.50	2.68	3.70
C.D.	00.43	5.14	5.62	6.10	6.40	6.90	9.50

Per cent growth inhibition is average of three replicate.

References

Bhale, U.N. and Gogle, D.P. (2008). Effect of passage on the development of carbendazim resistance in *Alternaria spinaciae* incitant of leaf spot of spinach (*Spinacea oleracea* L). *Geobios*, pp. 35-37.

Dekker, J. and Gielink, J. (1979). Acquired resistance to pimaracin in *Cladosporium cucumarinum* and *Fusarium oxysporum f.sp.narcissi* associated with decreased virulence. *Neth. J. Plant Pathol.*, **85**: 67-73.

Gangawane, L.V., Arora, A.K. and Kamble, S.S. (1993). Effect of passage on the development of metalaxyl resistance in Phytopthora infestans. In: *Chemical Management of Plant Pathogens in Western India,* (Eds) L.V. Gangawane, D.S. Mukadam, P.B. Papdiwal and S.R. Shinde, IPS (WZ). Publ, Aurangabad, pp. 27-30.

Griffin, M. J. (1981). *Plant Pathology*. Notes No. 38, Fungicide Resistance ADAS South Western Region, U.K.

Horsten, J.A.H.M. (1979). Acquired carbendazim resistance in *Septoria nodorum* and *Cercosporella herptotrichoide* in cereals. Dissertation Agricultural Univ. Wageningen, Netherland, pp. 107.

Hartill, W.F.T. (1979). Resistance of plant pathogens to fungicides. *Newzeland J. Except. Agric.* **14**: 239-245.

Shabi and Glpatric, J. D. (1981). Competition between benomyl resistant and sensitive strains of *Venturia inaequalis* on apple seedling treated with benomyl and capton. *Neth. J. Plant Pathol.*, **67**: 250-251.

Index

N

O

P

T